SAFETY AND LABORATORY PRACTICE

Macmillan Technician Series

P. Astley, *Engineering Drawing and Design II*

P. J. Avard and J. Cross, *Workshop Processes and Materials I*

G. D. Bishop, *Electronics II*

G. D. Bishop, *Electronics III*

J. Elliott, *Building Science and Materials*

D. E. Hewitt, *Engineering Science II*

P. R. Lancaster and D. Mitchell, *Mechanical Science III*

R. Lewis, *Physical Science I*

Noel M. Morris, *Electrical Principles II*

Noel M. Morris, *Electrical Principles III*

SAFETY AND LABORATORY PRACTICE

John G. Ellis

Norman J. Riches

M

First published 1978 by
THE MACMILLAN PRESS LTD
London and Basingstoke
Associated companies in Delhi Dublin
Hong Kong Johannesburg Lagos Melbourne
New York Singapore and Tokyo

Printed in Great Britain by A. Wheaton & Co., Ltd., Exeter

British Library Cataloguing in Publication Data

Ellis, John Graham
  Safety and laboratory practice. — (Macmillan
  technician series).
  1. Laboratories — Great Britain — Safety measures
  I. Title   II.   Riches, Norman J
  614.8′3   Q183.G7

  ISBN 0–333–23312–3

Contents

We also acknowledge the assistance of the following organisations who have allowed us to reproduce copyright material: Association for Science Education, Air Products Ltd, British Oxygen Co. Ltd, Chubb Fire Security Ltd, Griffin and George Ltd, Hopkin and Williams. Extracts from BS 5378 are reproduced by permission of the British Standards Institution, 2 Park Street, London WIA 2BS, from whom complete copies can be obtained.

JOHN G. ELLIS
NORMAN J. RICHES

Preface

Our aim in writing this book has been to provide a useful text for students on courses leading to the Technician Certificate of the Technician Education Council in Sector C; this includes courses in all the sciences, medical laboratory technology, plastics and polymers.

It has been stated that the content of the Level I Standard Unit, or something similar to it, must be studied by most of these students and their knowledge must be assessed. This does not mean that safety must be taught separately from the technical units; indeed, many colleges intend to integrate the teaching of safety and laboratory practice with other units where this is appropriate. We have borne this intention in mind while preparing this book.

In some places we have gone beyond the TEC Standard Unit (U76/001). For example, we have included a chapter on hazards in the biology laboratory, and in the appendix we have included data which we hope will be of permanent value to those who use this book. There will be some students following a TEC programme without regular attendance at college, and we hope that they will find what we have written is useful.

This book will also be of value to those studying the Standard Unit Introduction to Health and Safety at Work (U78/450), a free-standing unit which may be included in a wide variety of programmes.

We have attempted to bear in mind the needs of those technicians who are not following formal TEC courses but nevertheless need to know how to conduct a laboratory safely. There will be those starting work in laboratories somewhat later in life—school part-time laboratory assistants are an obvious group. Many technicians recruited to the fast-growing laboratories in the developing world should find that, with the exception of details of English law, everything we have said applies to them as well.

We would like to thank all those who have helped in many different ways in the writing of this book. We would particularly like to thank John Rowlinson and Geoff Jones, for their considerable help in the preparation of chapters 5 and 7, and Ian Bourne, for his assistance in the preparation of many of the photographs that are reproduced here.

1 Laboratory First-aid

Although laboratories are potentially dangerous places, the records of accidents that occur show that the number of serious injuries is relatively low. In schools, the gymnasium and work-shops have more accidents, and more injuries occur in corridors than in laboratories. However, science technicians are often called on to give assistance and, considering the hazards in laboratories, they should obviously be capable of dealing with the more likely incidents.

In order to be competent to give assistance, it is essential to receive practical training and to gain experience. The first-aid courses organised by the St John Ambulance Brigade, the British Red Cross Society and other similar organisations are most worth while; practical training in bandaging, etc., is given, together with the opportunity to benefit from the knowledge of experienced first-aiders. These certificates are normally valid for 3 years, after which a refresher course and re-examination are required. Having gained a certificate it is also necessary to gain experience; the symptoms observed with a real patient are often very different from those described so precisely in training, and the techniques of reassuring and caring for a real patient cannot be learnt in the classroom.

The main aim of first-aid treatment is to ensure that the condition of a patient does not worsen, and preferably improves, while awaiting the treatment necessary for his injuries. In minor cases this treatment may be given by the first-aider, but in more serious cases, professional attention will be needed. The first-aider must be able to distinguish between those situations that he can deal with, and those that he can't, and he must adapt his actions in particular circumstances so that he is always working in the best interests of his patient.

1.1 BLEEDING

The human body contains approximately one litre of blood for every ten kilograms of body weight (or about one pint for every stone, that is, 14 1b). The body can survive without ill effect the loss of a half litre or so, this being the quantity given by a blood donor at each blood-donation session, but such losses, when combined with the shock of an accident, are more serious. If serious bleeding is observed then stopping it must be the first priority.

Blood leaving the heart to go round the body travels in arteries, with the familiar beating motion observed in the pulse, due to the heart's pumping action; as the arteries get smaller, which they do the further away they are from the heart, this beating becomes less noticeable, until, in the small capillary blood vessels at the extremities of the body, it disappears. These capillaries recombine to form the veins which carry the blood, without beating, back to the heart. In addition to the beating, the bright red colour of arterial blood distinguishes it from the blood in veins which is more purple.

Bleeding from most wounds can be stopped by direct pressure on to the wounded area. This pressure may be applied at first by holding a dressing in place by hand, and then fastening it by bandage or adhesive tape. Wounds containing foreign bodies, especially if deep, for example, a piece of glass tubing in the hand,

are best treated by pressure around the wound and not by first removing the object. A suitable ring that can be used to apply this pressure can be made from a triangular bandage or even a handkerchief (figure 1.1).

Direct pressure on the wound may be inadequate to stop severe arterial bleeding, and if copious bright red bleeding in pulses is observed, the nearest 'pressure point' should be used. Pressure points are places where an artery is near to the surface of the body

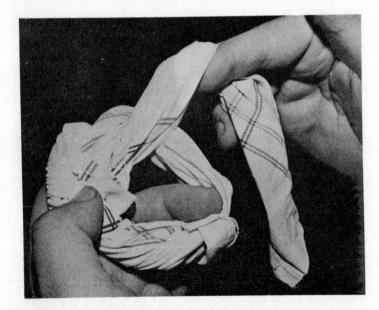

Figure 1.1   Ring bandage

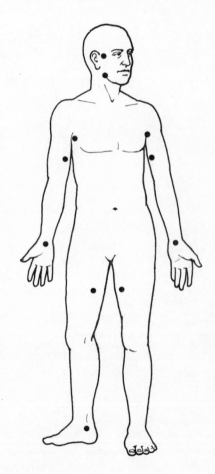

Figure 1.2   Main pressure points

with a bone lying just below. For example, arterial bleeding from the thumb can be controlled by applying pressure to the 'pulse' in the wrist, and the blood supply to the whole arm can be stopped by firm pressure on the inside of the upper arm, about half-way between shoulder and elbow. Figure 1.2 shows some of the main pressure points.

Under no circumstances should a tourniquet be used, nor should anything be tied tightly over a pressure point to stop bleeding. The first-aider himself should apply the minimum necessary pressure by hand.

Large open wounds may be plugged to stop bleeding. It is preferable that material used for this should be clean and sterile, but stopping serious bleeding is more important than the risk of infection.

Serious injuries may cause internal bleeding. This may be detected by the patient spitting up blood, or by a swelling, often dark in colour, near the injury. Such cases must be referred for medical attention as quickly as possible. The patient should be given nothing by mouth. There is no treatment for minor internal bleeding, such as bruises.

It is always helpful to raise the wounded part to reduce the blood pressure at that point. The blood stream, as one of its functions, carries the oxygen supply to the brain, and a serious blood loss will, therefore, deprive the brain of its necessary supply. This will lead in turn to dizziness, faintness, unconsciousness and death. Therefore the control of serious bleeding must always be a first priority.

1.2 FRACTURES

Some bones may become chipped or cracked, for example, in the fingers, or even broken, for example, in the wrist, without the patient feeling any more pain than that from a bruise, and without there being any obvious signs of a fracture. Fractures of the arms or legs are invariably painful and are accompanied by loss of power of movement; there will often also be a visible deformity or swelling.

If medical assistance or an efficient ambulance service is easily available, the first-aider should do nothing to treat the fracture. The patient should be reassured and kept as comfortable as possible with the minimum movement of the suspected fracture; if there are injuries to the spine the patient must not be moved without expert assistance unless there is an immediate threat to his life. Movement of any fracture may cause further injury and do more harm than good. The only cases where first-aid treatment of fractures should be attempted are in situations where skilled attention is not available, for example, following a major disaster or in accidents at remote field stations or on expeditions. In these latter cases, practical training and practice is necessary especially in immobilising a patient so that he may be safely transported on a rough journey.

1.3 HEAT BURNS

The treatment that is required *immediately* is to get as much cold water as possible on to the burnt area, by total immersion if possible. This cooling must be continued for up to fifteen or twenty minutes — time it, don't guess — in the case of severe burns, in an attempt to bring the temperature of the affected tissue back to normal. (The same treatment is effective for 'cold' burns; cold water is usually within about 25 °C of body temperature.) When this has been done satisfactorily, and after removing any rings, etc., that would constrict the swelling that is likely to follow, the burnt area should be covered with a clean dry dressing designed to exclude air and protect the wound from any further damage or infection. Burns *must* be referred for medical attention where the skin has been broken, or where the burnt area is larger than the palm of the patient's hand. For these burns, do not apply any ointments, creams, greasy dressings or cotton wool and do not attempt to remove any clothing or other material which may be adhering to the wound. Severe and large burns will lead to a major loss of fluid, usually blood plasma, and in such cases plain cold water in small quantities may be given to drink.

If clothes catch fire, the victim should be laid down and rolled in a carpet or blanket to extinguish the flames. This will prevent the flames being breathed in and causing damage to the lungs, etc. Do not remove burnt clothing. Then treat as for other burns.

1.4 CHEMICAL BURNS

Many chemicals are sufficiently reactive to cause damage to the skin. Commonly occurring examples are the caustic alkalis and the mineral acids, although phenol, chromic acid, hydrofluoric acid and phosphorus are not uncommon causes of burns. If a substance known to be corrosive is to be used, reference books should be consulted to find any specific antidotes, but in any cases where a specific antidote is not immediately available, the most satisfactory treatment is generous washing with clean cold water. Because of the long-term effects of many of the substances involved, victims of chemical burns should be referred to a hospital, giving as much information about the accident as possible.

1.5 POISONING

Many substances in common use in laboratories are labelled as poisonous, but it should be assumed that all laboratory reagents are harmful. Many poisons are best treated by specific remedies—cyanides are probably the best known of these. Before using any particularly dangerous substance, find out if there are specific antidotes, and ensure that they are prepared ready for use.

If the victim of poisoning is conscious, he should not be made to vomit if there are signs of burning of the lips or mouth or if the poison is a petroleum product. What was corrosive when swallowed will also be corrosive when vomited; the patient should be removed to hospital as quickly as possible. Petroleum products are most dangerous if they get into the lungs, which could happen following vomiting. For all other poisons, provided that the patient is conscious, he may be made to vomit; a most effective method is to put a finger (either his or yours) down his throat to depress the back of his tongue, but a drink of salt in water or mustard in water may also be effective. In all cases give the patient a generous drink of water or milk to dilute any remaining poison

and make him vomit again. After a further drink, send him to hospital with a sample of the vomited material. In all cases also send any information available concerning the nature of the poisonous material.

1.6 UNCONSCIOUSNESS

The treatment of the unconscious patient is the same whatever the cause of the unconsciousness. Special precautions will, of course, be necessary to remove the patient from the cause of the unconsciousness in such cases as gas poisoning or electric shock, remembering that thoughtless haste is more likely to lead to two victims rather than one recovery.

First ensure that the patient is breathing. The brain will be irretrievably damaged if it is without oxygen for more than three or four minutes, and breathing is essential to replace the oxygen in the blood. If breathing cannot be detected, it will be necessary to give artificial respiration, or ventilation. It is then also important to check that the patient's heart is beating; check the pulse in the wrist or neck, or listen, with your ear, to the patient's chest. If you believe that there is no heart beat, then heart massage should be attempted.

When the patient's breathing is satisfactory, he should be placed in the recovery position (see figure 1.3). Turn him on to his side, with the uppermost leg pulled forward and bent at the knee, the arm beneath the patient being carefully laid behind him, and the upper arm brought up close to his head. If his head and neck are bent slightly backwards, the forward position of the chin will ensure that his tongue does not obstruct the air passages in his throat. Make sure that you do not affect any other injuries that the patient might have. If he is on a bed or stretcher, it will help the

Figure 1.3   Recovery position

flow of blood to his brain if the foot end of the bed is slightly raised, and the risk of inhalation of vomit will also be reduced. You may cover the patient with a blanket, but do *not* apply heat, for example, with a hot water bottle. Do not give the patient anything to drink. Do not leave an unconscious patient unattended. All patients who have been unconscious must receive medical attention.

1.7 ARTIFICIAL RESPIRATION

The most effective method of artificial respiration, or ventilation, is mouth-to-mouth, in which the first-aider forces his exhaled breath into the patient's lungs. The first step is to ensure that the patient's airway is not obstructed. Do this by moving his head back as far as possible, and letting his lower jaw come forward (if the patient is lying on his back, this can be done by lifting his neck from underneath) which will ensure that his tongue will not fall back. Also check that his mouth is clear of obstructions such as false teeth. Then, as you breathe deeply but naturally at about 10 or 12 breaths per minute, breathe out into the patient by placing your mouth over his, remembering to cover or close his nose. When you breathe in, avoid the air that your patient is exhaling (it may be poisoned) and watch to see that his chest is contracting. If a movement of the patient's chest cannot be seen, then you are not assisting him; check carefully that the airway is clear, and that the air you are giving him by mouth is not coming out through his nose.

There is a risk that an unconscious patient could be the victim of poisoning and that the poison could affect the first-aider making mouth-to-mouth contact. This risk may be reduced by using a handkerchief, for example, around the patient's mouth, or by using one of the special 'airways' designed for this purpose.

There are a few situations where the mouth-to-mouth method is not possible. These are usually associated with severe facial injuries. There are several alternative methods available, but these, at best, only give about half the air volume to the patient and are much more likely to cause other injuries; they are also more tiring for the first-aider. In the Silvester method, the patient should be placed on his back with his shoulders raised by a pad so that his

head is tilted back and his airway is cleared. Kneel behind him with his head between your knees. Hold his wrists across the lower part of his chest and rock forward so that your weight forces the air out of his lungs. Then sit back on your heels, lifting the patient's arms up and back as far as you can. Repeat this action about 12 times a minute, checking from time to time that his airway remains clear.

It is important to start artificial respiration as soon as possible and to continue until the patient begins to breathe for himself, or until a doctor pronounces death; you should certainly continue for at least an hour if necessary.

1.8 HEART MASSAGE

Victims of electric shock or certain poisons or those who suffer a heart attack may experience heart stoppage. A check should also be made in all cases of breathing failure to discover whether or not the heart is still beating. Feel for a pulse, or listen carefully on the chest. If there is no sign of a heart beat, place the heel of the hand at the base of the sternum (chest bone), not on the ribs, and give a sharp blow. Follow this with a firm press about once per second, adjusting the pressure to the size of your patient. It must be realised that the chance of success is slight, but even a remote chance makes the effort worth while. You should learn, by practice on a dummy, how to combine cardiac massage with artificial respiration — *skilled instruction is essential*.

1.9 SHOCK

Shock will follow any injury and is likely to be most severe following serious bleeding or burning. The symptoms are shallow breathing, weak and rapid pulse, paleness of the skin and a cold sweat and low temperature, particularly at the extremities of arms and legs. It is essential to realise that this condition can itself be fatal, and patients should be transferred to medical care, together with as much information on the cause as possible, that is, time and description of the accident, amount of bleeding, rate of heart beat measured at regular intervals and patient's temperature taken at intervals. The first-aider should comfort and reassure the patient and keep him warm with a blanket (no hot water bottles or warming near a fire) laying him down with his legs raised if possible. The treatment of any obvious bleeding is, of course, of the first importance. Warm drinks of tea or water (never alcohol) may be given if there are no symptoms of serious injury and provided that the patient is conscious.

Note The over-all rules about giving drinks to a patient are as follows:

(1) never give a drink to an unconscious patient
(2) never give alcohol in any form
(3) (a) drinks of water, tea, milk, may be given to conscious victims of poisoning to dilute any poison remaining after vomiting
(b) drinks of water, tea, etc., may be given, in small quantities, to conscious burned casualties
(c) drinks of water, tea, etc., may be given, in small quantities, to patients showing symptoms of shock.

Any drinks that you give will delay by several hours the possibility of giving an anaesthetic and therefore may delay professional medical treatment. It is always better not to give the patient any drink — the first-aider will probably benefit more than the patient from a cup of tea.

1.10 ORDER OF PROCEDURE ON DISCOVERING A CASUALTY

If the casualty is unconscious

(1) Assess the cause of unconsciousness. Be particularly careful if poisonous gas or electric shock is suspected. For poisonous gases, either use breathing apparatus, or ensure plenty of fresh air; for electric shock, either turn the supply off, or remove the victim by using suitably insulated implements. Retain samples of any suspicious materials, bottles, etc., and of any vomited material.
(2) Check the patient's breathing. If he is not breathing, also check the heart beat. Give artificial respiration and cardiac massage if required. (If he is breathing, you can be sure that his heart is beating.)

For all patients, whether conscious or unconscious

(3) Assess the injuries received — remember that a conscious patient may be able to tell you. Burns and bleeding must be dealt with as speedily as possible. Unless the patient is in danger from staying where he is, do not move him if you suspect any broken bones.
(4) Treat for shock.

At the earliest opportunity, summon help by sending someone to call an ambulance or doctor, unless the injury is so trivial that only first-aid will be required to ensure a complete recovery.

1.11 PROVISION OF FIRST-AID EQUIPMENT

Simple first-aid boxes should be provided close to the places where accidents are likely to occur (figure 1.4). They should contain only such items as adhesive dressings, small quantities of lint, gauze, cotton wool and bandages, a few pain-relief pills such as paracetamol BP or soluble aspirin, safety pins, forceps and scissors. There should also be a book for recording incidents for which the box was used. A designated person must be responsible for ensuring that the contents of the box are replenished and kept in good order.

More sophisticated equipment and a more comprehensive range of bandages, etc., should be kept at a central location under the charge of a doctor, nurse or trained first-aider depending on the size of the institution. A rest area will be needed and the extra equipment should include antiseptics, burn dressings, triangular bandages, a thermometer and possibly a stretcher together with any special equipment likely to be needed because of the specialist nature of the work that is being undertaken.

Details of the provision of equipment and the personnel involved will vary widely and it is important that you find out what should happen at your place of work and at college.

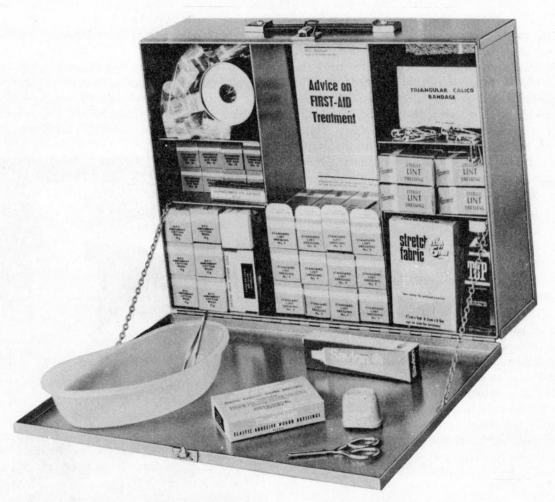

Figure 1.4   Typical wall-mounted first-aid box

1.12 DIABETES

Diabetes is the disease in which the production of insulin by the pancreas is defective. When the disease has been diagnosed it can usually be controlled by a correct diet and suitable daily injections of insulin. Provided that the diabetic follows the prescribed routine there is usually no problem. However, too little insulin will lead to too much sugar in the blood stream and the onset of a diabetic coma. More commonly, an excess of insulin, causing a shortage of blood sugar, will lead to an insulin coma. An insulin coma occurs comparatively rapidly, accompanied by shallow breathing and sweating. Fortunately, immediate treatment is simple: the diabetic should be given two lumps of sugar. Most diabetics carry some sugar with them for use in such a situation.

If the diabetic is already in a coma then, of course, he cannot be given anything by mouth. In the less likely event of a diabetic coma, the patient already has too much blood sugar, and giving sugar lumps cannot lead to a recovery. In either of these cases, medical assistance must be sought as soon as possible.

When diabetes has been diagnosed, the patient is invariably told how to control the disease. If you work with a diabetic, he will usually discuss with you what action to take if he feels that a diabetic or insulin coma may be starting.

1.13 FITS

Occasionally you may witness a person suffering a fit. This usually involves uncontrolled, powerful muscular movements with frothing at the mouth and rolling of the eyes. The first-aid treatment is to prevent the patient suffering further injury; this can be done by restraining the convulsions of the patient, especially in such places as laboratories, or, if possible, moving potential hazards out of the patient's reach. Observe as much about the behaviour of the patient as you can; which muscles were affected and how much, how long did the fit last, etc. Usually a fit will pass within a few minutes, and apart from exhaustion, the patient will return to normal. It is, however, important that the patient seeks advice from his doctor. It is unlikely that you will ever see unsuspected fits due to epilepsy.

1.14 EYE CARE

Many of the activities undertaken in laboratories present obvious hazards to the eyes. Chemicals can splash and particles from powered tools and abrasive wheels will fly considerable distances. There are specific regulations relating to many of these activities requiring the use of safety spectacles or goggles. However, many other activities can appear safe, but may easily lead to damage to the eyes should anything go wrong and if adequate eye protection is not being used.

Liquid splashes in the eyes should be washed out as speedily as possible with copious quantities of cold or tepid water. It is usually not worth while attempting to neutralise the contamination because this will delay treatment and is more hazardous than using plain water if done incorrectly. Take great care if you use the

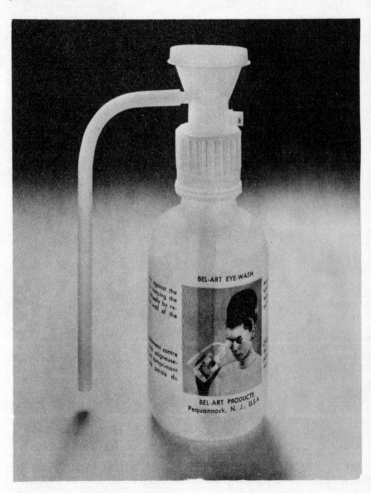

Figure 1.5   Eyewash bottle

normal laboratory cold water taps because these deliver a narrow, high-speed jet which can cause significant injuries. An eyewash bottle with a special cup attached will give a low-pressure water flow all round the affected eye. If these bottles are kept full of water, remember to wash out and refill them regularly to prevent the growth of algae. There is a variety of proprietary eyewash solutions available which may be more soothing than clean tap water, but will certainly be more expensive (figure 1.5).

Solid particles in the eye may be removed if they are visible near the eyelids or if they can be seen by lifting an eyelid away from the eye-ball. The affected eye will normally be watering which will help to float the particle out. Only use such soft implements as the corners of soft tissues or clean handkerchiefs to remove objects from eyes. If you cannot see the object complained of, if it has penetrated the eye-ball or is on the pupil or iris, cover the eye with a soft pad to prevent the victim rubbing his eye and send him to an eye hospital casualty department as quickly as possible.

1.15 REPORTING OF ACCIDENTS

It is essential to keep an accurate record of first-aid treatment given. This record should show

the date the treatment was given, and the date and time of the accident if different
the name of the patient, his age (accurately if under 18, approximately if older) and address
the injury suspected
the treatment given
any further action or treatment recommended
the name of the person giving the treatment.

The age of the patient may be particlarly important if questions involving the legal authority to give or to agree to receive treatment should arise. In the United Kingdom, a person over 16 may agree to his own treatment, but for younger people the agreement of a parent or guardian, or of a person temporarily in guardianship, such as a teacher for an accident involving a child at school, is necessary. The giving of treatment without proper consent could lead to a legal action for assault. Also, in all cases, the consequences of particular injuries will vary with the age of the patient, and the treatment must vary to allow for this.

The details of the treatment given and recommended are obviously important when investigating any mishaps, as well as when analysing the need for a greater or lesser provision of skilled first-aiders or first-aid facilities. There may also be claims for industrial injury benefit or for pay during sick leave and the patient may need to prove that he took all reasonable steps to ensure his own speedy recovery. Remember, however, that the first-aid records should not be confused with accident reports; we are not trying here to stop accidents happening or to assess the blame for those that do happen, but to record what was done to help the recovery of those involved. If people feel that their own possible carelessness will be noted and may be reported by the first-aider, they are much less likely to seek the treatment which is necessary.

It is, of course, also necessary to maintain an accident record. This should include a description of what happened and an attempt to state the cause of the accident, as well as the personal details of anyone injured and the names of witnesses. The accident record is required by law in certain work places, but should be maintained in *all* laboratories. If the record includes accidents which do not involve personal injury and also dangerous situations where accidents were narrowly averted, it will then be possible to use past experience to make the laboratory safer. All workers should be encouraged to view accident reporting this way and to give whatever information will be useful, rather than to attempt to hush-up unpleasant incidents for fear that they may be criticised or blamed.

1.16 ASSIGNMENTS

1.1 List the contents of the first-aid boxes nearest to your normal work place. Make sure that you know how to use everything there.

1.2 What is the procedure for dealing with casualties who need more attention than your laboratory first-aid box can provide?

1.3 Where are the nearest hospital casualty departments? Is there a special hospital for eye injuries?

1.4 How could you summon skilled help to a serious accident in your laboratory (a) at work; (b) at college? Does the procedure change outside normal working hours?

2 Electrical Safety

The coming of commercially distributed electricity has been one of the greatest benefits to modern society. Every laboratory in the country has a large number of electrically powered devices. Used carefully, electricity not only relieves the mechanical drudgery of many tasks but is indeed the very thing that makes many operations possible. Used carelessly, electricity can kill, not only through electric shock, but by causing fires. Frequently it is other people who suffer from an individual's carelessness. All technicians, no matter in what discipline they work, will be called on to use a variety of electrically powered equipment, and it is essential that they learn the fundamentals of safe working. Failure to do so makes them a danger, not only to themselves, but to the lives of others.

2.1 ELECTRICAL UNITS

It is not within the scope of this book to give an introduction to the basic principles of electrical theory. It is necessary, however, to explain the four basic quantities and the units in which they are measured — see table 2.1.

Table 2.1

Quantity	Name of Unit	Symbol
Current	ampere (amp)	A
Potential	volt	V
Resistance	ohm	Ω
Power	watt	W

2.1.1 Current

An electric current occurs when the free electrons in a conductor, for example, a piece of metal, start to drift along it. The passage of electrons through a piece of metal causes it to get hot and give out heat. In the case of a light bulb, the filament gets so hot that it glows white-hot and gives off visible light. The greater the flow of electrons, the greater the current is said to be (the word current is used in much the same way to describe the motion of water in a river). The unit in which the size of an electric current is measured is called the ampere (usually abbreviated to amp).

2.1.2 Potential

Potential is the name given to the force that is causing the electrons to drift, measured in volts. It is an electric potential that causes electric currents to flow; if there is no potential, there will be no current. The current will always flow from a point at high potential to a point at lower potential (in much the same way as rivers always flow from a high point to a lower one). In a normal mains socket in the United Kingdom the metal connections inside the lower left-hand hole are at a potential of 0 V and those inside the lower right-hand hole are at 240 V. If a material containing electrons that are free to move is connected between these two points, then an electric current will flow through that material.

2.1.3 Resistance

Resistance is the name given to the opposition that a material offers to the passage of an electric current. The unit of measurement of resistance is the ohm. A high-resistance connection to a mains socket as described above would allow only a small current to flow whereas a low-resistance connection would let a large current flow. A typical electric light bulb might have a resistance of about 1000 Ω, or 1 kΩ, and a one-bar electric fire a resistance of about 60 Ω. The resistance of the human body, which can also conduct electricity, depends on many factors such as perspiration on the skin, but is usually between about 500 and 10 000 Ω when measured from hand to hand.

2.1.4 Power

The power rating of a piece of electrical equipment measures the total amount of energy passing through that equipment in one second; it is measured in watts. A 1000 W, or 1 kW, electric heater will draw about 4 A from the 240 V mains supply, and a less powerful 500 W heater will draw about 2 A. In mains-operated equipment, the higher the power rating, the greater the current that will be drawn from the supply.

2.2 EFFECTS OF AN ELECTRIC CURRENT

The passage of an electric current through a material causes a variety of effects. A magnetic field will be created, a chemical change might take place and there will be a rise in temperature. If the current is passing through living tissue such as the human body there will be additional physiological effects. With very small currents a human victim will feel only a slight tingling sensation. At higher currents, the person's muscles will start to act involuntarily causing, for example, twitching or the violent movement of an arm, leg or even the entire body. Often, if a current is passing through someone because he has touched a 'live wire', these muscular contractions will cause the victim to tighten his grip on the wire and make it impossible for him to release his hold even though he knows he must do so. A person with an electric current passing between his hands only experiences a tingling numbness usually associated with a paralysis that lasts as long as the current is passing. Unless a person is assisted he will eventually die of asphyxia, since the respiratory system also suffers from this paralysis. At still larger values of current the heart will stop its normal pumping action leading to rapid death. (See also section 7.3.)

Apart from this physiological action, the passage of an electric current generates heat which will result in the victim being burnt particularly where he touches the live wire. The heat generated by the passage of an excessive electric current through a wire or flex can lead to fire with all its attendant dangers. Frequently, both in the home and the laboratory, cables are tucked out of sight under carpets, etc., and much wiring is, of course, under floorboards, in roof voids and cavity walls. A cable can overheat and start smouldering over a period of several months, remaining unnoticed until a fire actually starts.

There are, then, two aspects that must be considered when discussing the safety of electrical installations and equipment.

(1) Ensure that no one, as a result of using electrical equipment, is ever put in a situation where an electric current might pass through his body.

(2) Ensure that unnecessary electrical heating never occurs in wires, cables or flexes, plugs, sockets or other connectors.

2.3 CORRECT WIRING OF A 13 A PLUG TOP

Statistics suggest that at least two people die every year in the United Kingdom as a direct result of incorrectly wiring a plug top. The ability to do this correctly and efficiently is an essential skill for any technician. The following procedure is suggested.

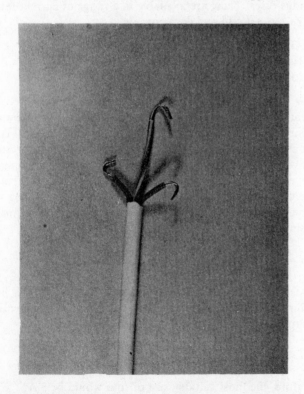

Figure 2.1   (a) Flex prepared for connecting to plug top

Carefully remove the outer insulation from the flex using a pair of wire strippers set to the correct diameter. Inspect the three insulated inner wires to see that their own (coloured) insulation is in no way damaged. If it is, cut the three loose ends off and start again. About three centimetres of the outer insulation should be removed in this way (figure 2.1a).

The three wires are colour coded for easy identification — see table 2.2.

Table 2.2

Wire	New Colour Code	Old Colour Code
Earth	Yellow and green stripes	Green
Live	Brown	Red
Neutral	Blue	Black

If necessary cut the live and neutral wires to the length required for them to reach their proper connections and remove the insulation from the end centimetre of each of the three wires. Tightly twist together each bundle of wires. The flex can now be connected to the plug top as shown in figure 2.1b, laying the wires around the terminal posts in a clockwise direction. Check that the metal contacts in the plug are touching the whole bundle of wires and not just a few of them, and that there are no loose strands of wire that are free to move about; if there are, carefully cut them off.

All plugs have some way of securing the flex so that the wires cannot be pulled away from the connections; check that this is secure. Fit an appropriate fuse and then replace the cover of the plug top.

2.4 CORRECT CHOICE OF FUSE

There are many different types and shapes of fuse (see figure 2.2), but they all work on the same principle. A gap is made in one of the conductors, usually the live one, and the fuse is fitted into this gap.

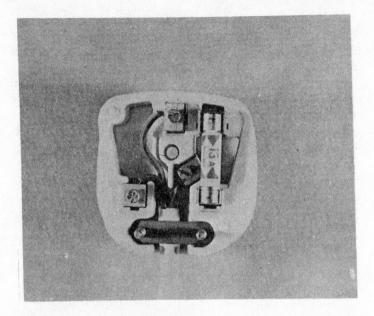

Figure 2.1   (b) 13 A plug top

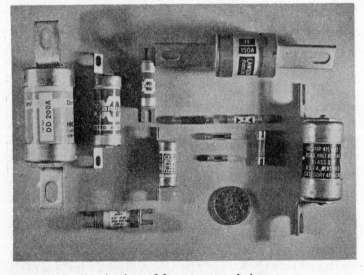

Figure 2.2   A selection of fuse types and sizes

When the circuit is operating normally the correct current flowing through the fuse will merely cause it to get warm. However, if there were a fault in the circuit which causes a larger current to flow, the fuse would get hot and melt, or 'blow' and therefore stop the current flowing altogether. The fuse functions as a simple safety device, allowing normal currents to flow uninterrupted, but if the current should rise above a predetermined value for any reason, the fuse automatically stops the current flowing. Commercially available fuses are specified by their current rating. A 10 A fuse, for example, will allow currents up to approximately 10 A to flow uninterrupted, but will melt and break the circuit should the current rise much above this value.

Fuses that are designed to fit plug tops are 25 mm (1 in.) long cartridge fuses. These are available in a range of current ratings, but the most useful and common are

3 A	Colour coded	Red
10 A	Colour coded	Black
13 A	Colour coded	Brown

It is important to fit a fuse capable of carrying the necessary current for normal operating conditions, but which will blow if the current rises substantially above this value. All electrical equipment should carry a small plate which states the manufacturer's model or serial number, the potential for which the equipment is designed, for example, 240 V, and the power rating of the equipment, for example, 480 W. It is possible to determine a value for the current required by this piece of equipment in normal use by using the formula

$$I = \frac{W}{V} \tag{2.1}$$

where I is the current to be calculated, W is the power consumption of the equipment in watts, and V is the potential of the supply. In the above example, the current would be

$$I = \frac{W}{V} = \frac{480}{240} = 2 \text{ A}$$

Therefore, the most suitable size of fuse would be 3 A.

It should be noted that equation 2.1 may give incorrect results if the circuit contains circuit elements which have properties other than pure resistance.

The approximate current drawn by, and the most suitable fuse size for some items of electrical equipment, *assuming a supply potential of 240 V*. are listed in table 2.3.

Table 2.3

Power Rating	Approximate Current	Fuse
100 W	0.4 A	
250 W	1 A	3 A
500 W	2 A	
1 kW	4 A	
2 kW	8 A	10 A
3 kW	12 A	13 A

Frequently it will be found that plugs that are purchased fitted with a 13 A fuse have been connected to equipment with a low power rating. This is a practice to be deplored, and it is essential that a fuse of the correct rating be substituted. If it is not, then there is a possibility that a current of 10 or 12 A may be passed by a wire that was only designed for a maximum current of 2 A. If this happens, the wire will overheat with the consequent risks of fire. If a 3 A fuse had been fitted, then it would have blown when the current exceeded 3 A thus cutting the current off before it had any chance of causing overheating and, by the equipment not functioning, would have warned the operator that there was a fault.

If a fuse of the correct size continues to blow, it is a sure sign that something is wrong and the fault must be attended to. Do not replace the fuse with a bigger one, even as a temporary measure.

2.5 THE EARTH WIRE AND EARTHING

The earth, particularly if it is damp, conducts electricity extremely well. At local substations, the electricity board connect the neutral wire of their supply to earth. (In its simplest sense, an earth connection of this nature is simply a copper pipe driven several metres into the ground with the neutral supply wire electrically connected to this pipe.)

Consider a small piece of equipment that is connected to the mains supply as shown in figure 2.3; due to an oversight, only the live and neutral wires have been connected.

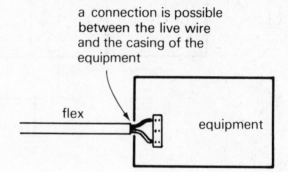

Figure 2.3

It may well happen that the casing of the equipment is made of metal. In time, the equipment may be moved about a lot and it is possible that the insulation of the live wire may become frayed or cracked and a bare live wire might come into contact with the metal casing. If this happens the metal casing will become live, that is, it will be at a potential of about 240 V. The neutral wire is at a potential of 0 V and so is the earth because they are connected at the substation. The situation now is shown in figure 2.4.

If an operator touches the live casing of the equipment, then his hand will be at a potential of 240 V and his feet, which are connected to earth through the building, will be at a lower potential. A current will therefore flow through his body from hand to foot. It is possible that this current could be very small because the building, the soles of the operator's shoes, etc., will have a large resistance and this will limit the size of the current. For an operator wearing rubber-soled shoes, it is likely that the current will be sufficiently small to be undetectable. However, suppose that with his other hand the operator touches a metal gas or water

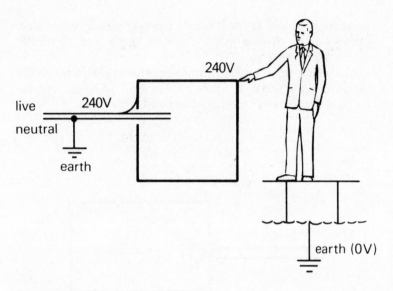

Figure 2.4

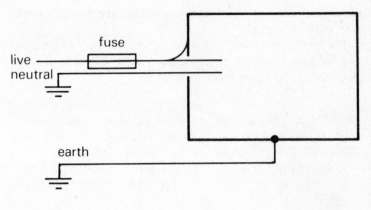

Figure 2.5

pipe which is buried deep in the ground and is at 0 V. The building, the operator's shoes, etc., are no longer in the circuit and the only resistance between the 240 V of the live casing and the 0 V of the metal tap is that of his body. If on this occasion, his resistance is 10 kΩ a current of about 25 mA will flow from one hand to the other. This is quite sufficient to cause the muscular contractions mentioned previously and put the operator at grave risk.

In order to overcome this problem, all mains sockets have three connections, the third being connected directly to an earth (figure 2.5). This is connected, by the yellow and green earth wire, to the metal casing of whatever apparatus is in use. If the casing becomes live then a current will flow from it at 240 V to earth at 0 V, but now, instead of the circuit incorporating the relatively high resistance of the operator's body, there is a thick metal wire connection which has a very low resistance. The current that will flow will be enormous, but it will not last for long because the protecting fuse will blow almost instantly thus disconnecting the supply and removing the high voltage from the casing and safeguarding the operator.

It can be seen that in normal operating circumstances, the earth wire has nothing to do — the equipment will operate perfectly satisfactorily without it. Because there would be no noticeable effect if the earth wire became accidentally disconnected and the equipment would continue to operate, it is important that the condition and the effectiveness of this earth connection should be tested periodically. The wire might have nothing to do under normal operating conditions, but if a fault does develop, then it can save life.

Inside the equipment, the earth wire should be fixed firmly to the metal frame or case. A soldered joint is not sufficient, since these can pull away after a period of time; preferably a combination of mechanical fixing and soldering should be used.

Inside the plug top, the earth wire should make good contact with the earth pin. Remember that the object of the earth wire is to carry a larger current than the rating of the protective fuse. If the earth connection is made only by a single strand of wire, then this may well melt before the main fuse blows thus disconnecting the earth.

There are several commercial devices available for testing the quality of the earth connection on a piece of equipment.* Usually a power supply is connected so that one side of it (0 V) is connected

* For example, the Test Set manufactured by The Clare Instrument Co., Worthing, Sussex.

to the earth pin of the plug and the other side (30 V approximately) is connected to the metal case of the equipment under test, an ammeter reading up to a minimum of 10 A being included in the circuit (figure 2.6). The power supply is switched on and the ammeter response noted; a steady large current should be indicated showing that the earth connections are capable of carrying currents of at least 10 A. If the current rises and then falls to zero, it indicates that there was a poor earth connection which has melted or otherwise failed. If the meter reading fails to rise above an amp or two, it implies that for some reason the earth connections have too high a resistance, which makes them quite useless. In the last two cases, checks should be made to determine the source of the fault which must be rectified before that equipment is used again.

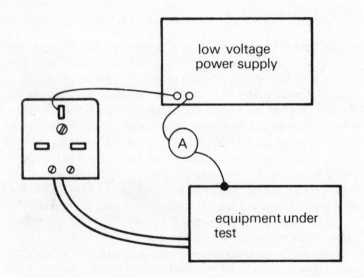

Figure 2.6

2.6 REPLACING EQUIPMENT FLEX

As has already been stated, all electrical wiring has a stated maximum current capacity. If a larger current is made to pass

through the wire it will overheat, which will result in the insulation becoming less effective, thus increasing the chances of live wires coming into contact with metal casings, etc. There is also the possibility of such heated wires starting a fire.

In general terms, the thicker the conductors in a piece of cable or flex, the larger its current-carrying capacity, but also the greater its cost. It is important, therefore, to use a flex which is sufficiently large to carry the current that is required but which is also the most economical for the job.

Manufacturers' specifications for cables usually quote the type of cable under a general heading and then specify the dimensions of the current-carrying metallic conductors. For example

flexible mains cables three core
$3 \times 19/0.1$ copper — PVC outer sheath diameter 3.7 mm

This refers to three-core cable, each core consisting of 19 strands of copper each one of which is 0.1 mm in diameter. The whole flex is made up in an outer PVC covering whose diameter is 3.7 mm.

three-core flexible mains cables

$3 \times 19/0.1$ copper	1 A	outer diameter 3.7 mm
$3 \times 13/0.2$ copper	2 A	outer diameter 5.0 mm
$3 \times 24/0.2$ copper	6 A	outer diameter 6.9 mm
$3 \times 40/0.2$ copper	13 A	outer diameter 7.5 mm

2.7 USE OF DISTRIBUTION PANELS AND MULTIPLE ADAPTERS

Provided that careful attention is given to protecting connecting flexes with an appropriate fuse, there is no danger in using multiple adapters of the type that allow you to plug several appliances into a single outlet. Consider, for example, the case of a distribution panel having five outlets which is connected to a single 13 A socket. It is connected by a three-core flex that is rated at 10 A and the plug attached to this flex has a 10 A fuse fitted (figure 2.7). Connected to the five outlets of the board are pieces of equipment which are rated at

1 kW, 500 W, 280 W, 700 W, 400 W

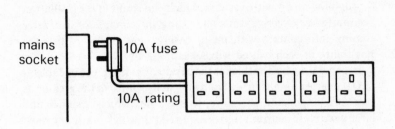

Figure 2.7

that is, there is a total power consumption of 2880 W. The current that would be drawn into the distribution board can be calculated approximately using equation 2.1

$$I = \frac{W}{V} = \frac{2880}{240} = 12\ \text{A}$$

As can be seen, the equipment is trying to draw a current of 12 A through a cable rated at only 10 A. If a 13 A fuse had been fitted, then this might have caused the cable to overheat, but since a fuse has been fitted that was compatible with the rating of the cable, that is, 10 A, this will blow indicating that the user was trying to 'overrun' the panel.

2.8 MECHANICAL DAMAGE TO ELECTRICAL FITTINGS

Even with taking normal care, plugs, sockets, etc., may become chipped or cracked. It is essential that such damaged fittings are replaced promptly.

After several years' use plugs or sockets may appear blackened around one of the terminals. This is probably due to a poor contact internally that is giving rise to arcing or sparking and consequently the generation of heat. Such sockets must be replaced immediately they are noticed; quite apart from the potential fire risk due to overheating, in a laboratory where there may sometimes be flammable vapours present, the risk of explosion is obvious.

Mains leads where the insulation is cracked or hardened (or frayed in the case of cotton-covered flex) must be replaced at once because they expose anyone using them to the risk of coming into contact with a live wire.

Mains leads must never be carelessly routed. They must not be allowed to drape across gangways, lie across a floor that is in use or be anywhere that might cause someone to trip. There is the obvious danger of a piece of experimental equipment being pulled on to the floor and breaking, perhaps taking other things with it which could be harmful to the person who has tripped, quite apart from the risk of injury due to falling; also a mains lead not securely fastened to the equipment might be pulled out leaving a live wire loose on the floor.

2.9 THE EURO PLUG

At the time of writing there is a suggestion that Europe, and possibly countries outside Europe, could standardise on a new type of plug and socket. Instead of having a maximum rating of 13 A as at present, this new system would have a maximum rating of 16 A. Rather than the pins being arranged in a triangular pattern, they would be almost in a straight line with the centre earth pin being slightly offset to prevent the plug being inserted upside down. These new plugs, which would not contain fuses, would be much slimmer than the existing 13 A type and a new double socket would take up as much space as a 13 A socket does at present. If such a system were introduced, it is likely that standards would be reorganised so that the plugs could be permanently moulded on to a flex end.

Such a new system has many advantages and disadvantages, but even if it were introduced into the United Kingdom now, it seems unlikely that the present 13 A system would be completely replaced for 20 years or more.

2.10 EARTH LEAKAGE TRIPS

In addition to fuses, or occasionally instead of them, many

establishments use earth leakage trips as an electrical safety measure. Although there are many different types available, in essence they all perform a similar function in that they sense the presence of a current in the earth circuit and then, by means of a relay, switch off the supply of current. They can be very sensitive, and can be made to switch off the mains supply if there is as little as 5 mA in the circuit. It can be seen that in many respects this provides a greater measure of safety than a fuse, where a current greater than the rating of the fuse is required in order to break the circuit. Since the smallest mains fuse in common laboratory use is 3 A, the extra measure of protection offered by a leakage trip can easily be seen.

If a leakage trip keeps operating and cutting off the supply, under no circumstances attempt to 'jam' it in, but try to ascertain what fault is causing it. As a first step switch off all the electrical equipment in the room and test each piece for high resistance between the live pin of the plug top and the earth pin.

Many items of equipment have neon lamps to indicate 'mains on'. Frequently the wiring is such that each neon lamp is causing a very small current in the earth circuit, and if there are many such lamps, the resultant current may be sufficient to cause the leakage trip to operate. It will then be necessary to rewire the equipment, with the indicators connected between live and neutral instead of between live and earth.

2.11 ASSIGNMENTS

2.1 Check which size fuse has been fitted in the plug attached to various pieces of electrical equipment both at home and at work where this is convenient. In each case, confirm that the fuse is of the correct size.

2.2 For several pieces of equipment, confirm that the earth connection is adequate both from an electrical and mechanical point of view.

2.3 Find out if the laboratory in which you work has a simple means whereby the mains electricity supply can be shut off quickly and totally in an emergency.

2.4 Imagine that a colleague has been electrocuted: list, in order, the steps you would take to help him.

3 Fire Safety

3.1 FIRE HAZARDS

Fire damage is now a very expensive matter as well as being one direct cause of injury and death. During 1975, for example, the cost of fire damage in the United Kingdom has been estimated at more than £4 for every man, woman and child in the population. It is therefore very important to do everything possible to reduce the risk of fire, especially in the high-risk areas such as laboratories.

A fire will occur when a combustible substance is in the presence of a supply of oxygen at a temperature which is sufficiently high. The fire usually gives out enough heat to maintain this temperature and keep itself burning. This may be represented by figure 3.1 — the triangle of fire, which shows the three necessary components of any fire.

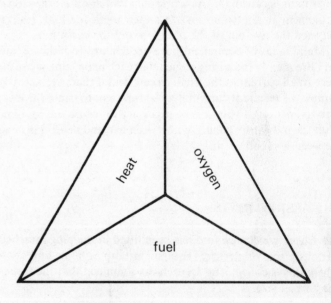

Figure 3.1   The triangle of fire

The combustible substances are also called flammable (the term 'inflammable' is falling out of use because 'in-' usually means *not*). Common laboratory materials which are flammable include

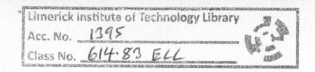

Solids
>   sodium, potassium, magnesium and naphthalene, as well as the carbonaceous materials such as wood, paper, rags, etc.

Liquids
>   ether, alcohols, carbon disulphide, petroleum and many common solvents; spirit duplicator fluid is also flammable

Gases
>   hydrogen, natural gas, town gas and the vapours of flammable liquids.

Some plastics materials burn fiercely, often giving off dense, acrid and maybe poisonous smoke, but other plastics are available which only burn slowly or even not at all. Polystyrene ceiling tiles, for example, if they have been painted with an oil-based paint, will burn and melt, falling down to the floor in burning sticky globules. The danger to people is obvious and there has been nationwide publicity about the dangers of painting such tiles, particularly in the kitchen areas of private houses.

An additional danger from burning liquids is due to the possibility of a fire spreading as the liquid flows away from the source of the fire. This may be helped either by extra liquid from a broken container or by water mistakenly used in an attempt to extinguish the blaze; many flammable liquids float on water, and the use of water on such a fire merely serves to spread it. It is important that any storeroom or cupboard used for keeping flammable liquids should be constructed so as to be able to contain the total volume of the stored liquids without any leakage. Either the floor should be below the level of the surrounding floors, or there should be a coaming at the bottom of the door (as in ship doorways).

Flammable vapours are often denser than air and will therefore flow and pour rather like liquids. As a result of this a layer of vapour may pour off a bench and flow across the floor to be ignited by an electric fire, cigarette end or spark from an electric switch which may be a considerable distance away from the original source of the liquid (figure 3.2).

A particular incident which comes to mind is that of a technician

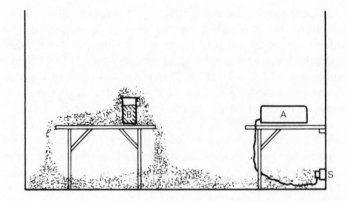

Figure 3.2   As the liquid in the beaker, B, evaporates, heavy vapour may fall down and spread across the floor where it may be ignited by a spark at the socket, S

carrying a partially full bottle of ether. The shaking around and the warmth from his hand combined to raise the pressure of the ether vapour inside the bottle. Eventually the internal pressure was sufficiently high to lift the glass stopper slightly and ether vapour flowed out and down on to the bench where an electrical heating mantle was switched on. The three basic constituents of a fire were then present: the flammable ether vapour, the oxygen in the atmosphere and the heat from the mantle. The ether vapour caught fire and 'flashed back' to the bottle, blowing out the stopper and leaving the technician holding a bottle of blazing ether. His reaction was, not surprisingly, to drop the bottle. It shattered when it hit the floor and a pool of fire began to spread from the centre of the accident.

An explosion occurs when the speed of a flame going through a flammable material is excessively rapid. This occurs in certain mixtures of solids, but particularly with gaseous mixtures of air and flammable material; note that many mixtures of dusts with air are highly explosive. (Coal dust and flour are two examples but, surprisingly, even stone dust can form an explosive mixture in this way.) For each mixture there is a precise concentration at which it is most explosive; more air or less air will make the mixture less explosive.

As mentioned previously, burning requires oxygen. Most commonly this is supplied by the atmosphere which usually contains about 20 per cent oxygen. There are, however, other important sources of supply. Most oxidising agents will support combustion and it is essential that laboratory oxidising agents such as peroxides and nitrates (including nitric acid) should be stored separately from flammable materials.

A fire will not occur unless both the flammable material and the oxidising agent are at a sufficiently high temperature. This temperature obviously may be caused by a flame but other causes include sparks from electrical equipment or a machining process; the heat generated by friction in badly lubricated bearings such as those found in motors, or radiant heat from the sun may start fires. Sparks from electricity generated by friction, so-called static electricity, can be a cause of fire; when rubber, nylon or other synthetics which are insulators slide against each other or against fixed objects very high voltage charges may be built up and sparks will occur unless adequate precautions have been taken. Such precautions include proper earthing connections or the application of conducting coatings, as with the anti-static cloths used on records; a damp atmosphere will restrict the growth of high voltages. Even nylon laboratory coats or oil flowing through pipelines can generate dangerously high voltages. The so-called 'spontaneous combustion' of haystacks or piles of damp waste is due to the heat from internal microbiological action. Such piles of flammable material are good thermal insulators and the heat generated is unable to escape, which can result in a dangerous rise in the internal temperature.

The _flash point_ of a liquid is the lowest temperature at which it gives off sufficient flammable vapour near its surface for a fire to occur, provided that air and a suitable source of ignition are present. It will be clear that low flash points represent a high fire risk. It will be safer to store flammable liquids at low temperatures to reduce evaporation and the production of explosive vapours. However, domestic refrigerators should not be used for this purpose because their thermostats and interior-light switches are not 'flame-proof'; that is to say, vapours may penetrate the thermostats or switches and be ignited as the electrical contacts open.

The _auto-ignition_ or _self-ignition temperature_ is the temperature at which a substance will ignite itself in air without being set alight. The presence of extra oxygen in the atmosphere will usually increase the fire risk by lowering the auto-ignition temperature.

Table 3.1   Fire Hazard Details for Some Common Laboratory Substances

Substance	Flash Point (°C)	Auto-ignition Temperature (°C)	Explosive Limits in Air (% by Volume)	
			Lower	Upper
Acetaldehyde	−38	185	4	58
Acetic acid	43	426	4	17
Acetone	−18	538	3	13
Acetyl chloride	4	390	—	—
Amyl alcohol	20 to 45	300 to 450	1	10
Benzene	−11	562	1	5
Butanol	25 to 30	360 to 400	1	4
Carbon disulphide	−30	100	1	45
Diethyl ether	−45	180	1	9
Ethanol	12	423	3	3
Ethyl acetate	−4	427	2	6
Methanol	10	464	7	37
Petroleum spirit	−17	288	1	6
Styrene	31	490	1	6
Toluene	4	506	1	7
Xylene	17 to 25	460 to 530	1	7

3.2 FIRE EXTINGUISHERS

Most fire extinguishers depend for their action on either lowering the temperature of the fire below that at which burning will continue, or removing the oxygen supply from the fire. There are, however, some types of extinguisher that work by interfering with the chemical processes occurring within the fire.

The European Standard EN 2 (which has superseded BS 4547) classifies fires into four types depending on the flammable materials involved.

Class A   Fires involving carbonaceous materials; this includes wood, paper, furniture, etc.
Class B   Fires involving flammable liquids
Class C   Fires involving flammable gases
Class D   Fires involving flammable metals

It is important to remember that any of these classes may also involve an electrical hazard due, for example, to the involvement of mains-connected equipment.

A major difficulty facing the fire fighter is the lack of any universally accepted colour code to indicate different types of fire extinguisher, nor does the shape or size of an extinguisher indicate its contents or mode of operation. Figures 3.3a–d illustrate some of the common shapes. Many contain water (soda–acid, water–carbon dioxide or water under stored pressure) or foam, and are commonly found in the shapes illustrated in figure 3.3a. Some dry powder extinguishers are also cylindrical, but they usually have conical sprays on the end of the hose. The method of operation of this type of extinguisher varies and technicians should familiarise themselves with the instructions printed on those in their own establishment.

The type of extinguisher shown in figure 3.3b gives off carbon dioxide gas at high pressure; some dry powder extinguishers may look similar although they have much thinner-walled containers. All these are operated by removing a safety pin and squeezing a trigger.

The types shown in figures 3.3c and d usually contain one of the heavy vapourising liquids. The one illustrated in figure 3.3c usually requires pumping whereas that in figure 3.3d operates in a very similar manner to the carbon dioxide extinguisher mentioned previously.

3.2.1 Main Types of Fire Extinguisher

Water

These may contain tap water and a small carbon dioxide cartridge which can be pierced to give the pressure necessary to expel the water as a jet (like a soda siphon) and are referred to as water–carbon dioxide extinguishers. Others are directly pressurised when filled and referred to as stored pressure extinguishers. A third type relies on mixing an acid, usually sulphuric, with an alkaline solution, usually of sodium carbonate. The chemical reaction between these two solutions generates carbon dioxide, and it is the pressure created by the production of this gas that drives the liquid out in a jet.

Effect   This type of extinguisher works mainly by cooling the flammable substances although the steam that is generated does tend to exclude air from the fire.

Good   for class A fires (carbonaceous materials).

Dangerous   for class B fires (burning liquids) and any fire that involves an electrical hazard.

Comments   Older extinguishers of this type cannot be stopped and restarted, but they are fairly cheap. They usually contain about 9 litres (2 gallons) and have a mass of about 13 kg (30 lb). The soda–acid versions must be refilled every three or four years even if they are unused.

Foam

The pressure required to expel the foam from this type of extinguisher is generated in the same manner as in the soda–acid water extinguisher described above. In addition, the extinguisher contains gelatinous materials so that the bubbles of gas form a sticky foam.

Effect   This type works by excluding air from the site of the fire. An unbroken blanket of foam needs to be laid over the fire, and any breaks that may occur in this blanket must be refilled.

Good   for class B fires (burning liquids).

Dangerous   for fires with an electrical risk.

Comments   The comments for the water extinguishers also apply

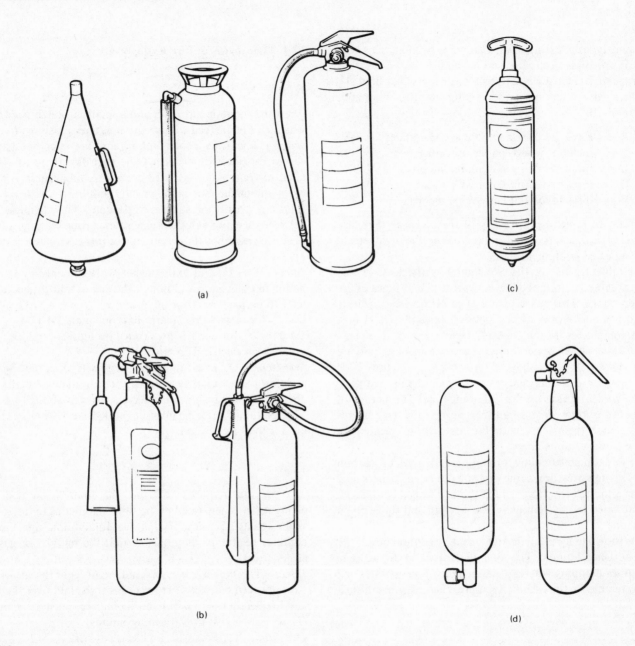

(a)

(b)

(c)

(d)

Figure 3.3

to the foam type. However, much more cleaning up is required after the fire has been put out and this type is not recommended for use indoors.

Heavy Vapourising Liquids

These extinguishers are usually of the stored pressure type, operated by squeezing a trigger, although some older versions incorporate a pump. (Small aerosol versions are also available, but these are only suitable for the smallest fires.)

Effect This type of extinguisher works by excluding air from the fire by forming a layer of dense vapour over the area and in some cases can interfere chemically with the reaction of the fire.

Note Older types of these extinguishers contained carbon tetrachloride, methyl iodide or methyl bromide. These compounds are themselves poisonous and carbon tetrachloride can break down to give phosgene, the First World War poison gas. Newer extinguishers use bromochloro difluoromethane (known as BCF) or Freon. These are much less poisonous than the chemicals used in the older types.

Good for class B fires, and are safe for fires involving an electrical risk. They are also very suitable for fires in electro-mechanical equipment such as tape recorders because there is no mess to be cleared up once the fire is extinguished.

Dangerous if used in a severely confined space, but the smell they leave after a fire is so pungent that this is not a serious risk with the newer types provided that an exit is available.

Comments This type is rather expensive by comparison with foam, but is easier to use and leaves no residue. This is particularly useful with fires involving equipment such as tape recorders or projectors, because no damage will be done beyond that caused by the fire.

Carbon Dioxide

These extinguishers consist of a small cylinder containing carbon dioxide at high pressure. This gas will be released, usually very noisily, when a trigger is squeezed.

Effect This type of extinguisher works by excluding air from the fire but carbon dioxide, also more dense than air, is a relatively light gas which will blow away quite easily.

Good for class B fires, fires involving electrical risks and fires in electro-mechanical equipment.

Dangerous if used in confined spaces; carbon dioxide suffocates people at lower concentrations than those required to extinguish fires; this is the opposite of the effect of BCF.

Comments Due to the high pressures involved, extinguisher cases are thick and quite heavy, but the gas is cheap. These extinguishers are very good for fixed, automatic extinguishing systems. When used at close range, this type of extinguisher can exert considerable force and care should be taken not to create a situation equally as hazardous as the original fire by knocking over reagent bottles, cultures, etc.

Dry Powder

In this type of extinguisher, a finely divided powder, usually consisting mainly of a bicarbonate, is blown on to the fire by gas pressure. The heat of the fire liberates carbon dioxide from the bicarbonate actually at the fire.

Effect The carbon dioxide that is liberated at the fire excludes oxygen.

Good for class B fires, but there are general-purpose powders that are also suitable for class A fires.

Dangerous No risks involved.

Comments These extinguishers are fairly cheap to refill. They are particularly effective on oil fires.

Fire Blankets

These were traditionally made of asbestos but modern alternatives such as glass fibre are now available. They are very suitable for dealing with personal clothing on fire and for limited quantities of burning material such as might be found in a small bench experiment.

Effect They exclude air from the fire.

Sand Buckets

These are very suitable for absorbing burning liquids that might spread. They can also deal effectively with class D fires by excluding oxygen.

We can summarise the suitability of extinguishers for the different classes of fire as follows.

Class A (carbonaceous)
Use water hoses, water extinguishers, carbon dioxide, general-purpose dry powder or a fire blanket. Do not use water if there is an electrical risk involved.

Class B (flammable liquids)
Use foam, heavy vapourising liquids, carbon dioxide, dry powder, a fire blanket or sand. Do not use foam if there is an electrical risk.

Class C (flammable gases)
Turn off the gas supply, then extinguish any remaining fire.

Class D (flammable metals)
Use sand or a fire blanket. For certain known risks, specially designed extinguishers may be available.

Table 3.2 The Suitability of Extinguishers for Different Types of Fire

| | Class of Fire | | If there is also an Electrical Hazard |
Extinguisher	A	B	
Water	√√	NO	NO
Foam	√	√√	NO
BCF	√	√√	√
Carbon dioxide	√	√	√
Dry powder	√*	√√	√
Fire blanket	√	√	√
Dry sand	√	√	√

* A special 'General Purpose' powder is available.

It may be very difficult to distinguish the type of extinguisher merely by noting its shape. The extinguishers produced by one major manufacturer differ significantly only in colour, the code being red for water, white for foam, black for carbon dioxide, green for BCF and blue for dry powder (see figure 3.4).

Figure 3.4

HAZCHEM code can be used to mark road vehicles, chemical stores or individual containers to indicate the proper emergency procedures. There may be better and different procedures for use in the laboratory where a wide range of materials suitable for specific antidotes is available. For example, a laboratory spillage of sulphuric acid would be dealt with best by neutralisation with

Explosive

Toxic

CORROSIVE
black on white

POISON
black on white

Oxidising agent

Corrosive

EXPLOSIVE
black on orange

FLAMMABLE SOLID
black on red and white stripes

Flammable

Harmful or irritant

FLAMMABLE GAS
black on red

OXIDIZER
black on yellow

Figure 3.5

an alkali, whereas the HAZCHEM code for sulphuric acid is 2P.

In order to give information to the public of the hazard involved, but not the emergency procedure, warning diamonds are being used increasingly. The EEC have introduced a series of chemical hazard warning symbols. Examples of both of these systems are given in figure 3.5.

3.6 ASSIGNMENTS

3.1 What types of fire extinguisher are provided at your place of work and at college? How does each type operate? Who is responsible for their maintenance?

3.2 At work, are the extinguishers conveniently placed, and are the correct types available at each location?

3.3 Are fire exits and the routes to them clearly marked? What sort of smoke doors and fire-resisting doors are provided? Are they kept closed?

3.4 At work, what warning systems are available for the occupants and how would the fire brigade be called? Do you have any specially trained staff to give 'fire first-aid'?

3.5 Where you work, are special hazards marked and are the markings accurate and adequate?

4 Technicians and the Law

4.1 THE ENGLISH LEGAL SYSTEM

The system of law in England, and in many other countries whose legal systems are based on the English system, has two main sources: case law and statute law. Case law, which may be described as judge-made law, is the source of the general rights to liberty and justice, and has developed over the centuries as judges have been called upon to decide what was fair in particular circumstances. As society has become more complex, it has become increasingly necessary to formalise and to change the rules of what is fair and to define limits on the rights and duties of individuals and organisations, and to spell out these rights and duties in a statute, or Act of Parliament. A recent example of this change is the Health and Safety at Work, etc., Act, 1974 which states, among other things, the duty of working safely to avoid harm to others, which supplements the common law* duty to show reasonable care for others.

There will be some instances where the meaning of the statute is not absolutely clear, even to lawyers. These instances often become the subject of test cases, in which a particular case goes before the courts to be decided, usually by High Court judges; because these cases will then become part of case law and will be used to decide similar instances in the future test cases are often supported financially by trades unions or associations who will engage the best lawyers in an attempt to obtain the interpretation of the law that they want. Many cases of this kind have concerned taxation, insurance and liability (responsibility) for accidents.

We can consider the law and its requirements as they relate to the technician as a worker, similar to other workers, to the technician as a person having special technical knowledge and responsibilities, and to the technician as he is involved in buying, and occasionally selling, equipment and materials.

Many statutes or Acts of Parliament include specific requirements, for example, the Customs and Excise Act, 1952 requires, in section 115, that the users of propyl, butyl, amyl or any other isomeric form of alcohol must keep a record of supplies, etc. Many other Acts give the power, usually to a government minister,

---

* that which has evolved from the rules of custom

to make regulations; the Health and Safety at Work Act, sections 15 to 17, authorises the establishment of the Health and Safety Commission which is given the power to prepare and enforce Regulations and Codes of Practice. Other Acts, such as the London Government Act, 1963, include certain requirements which affect laboratory work in a particular district. These Acts are part of the general law of the land, and failure to do as the law requires may result in prosecution in a court of law, with the possibility of a fine or even imprisonment. These prosecutions will be made by the police, or by inspectors appointed under the Health and Safety at Work Act, or by officials concerned with breaches of customs regulations, etc.

In addition to the law of the land, there may also be regulations made by an employer. These may cover such things as the type of protective clothing to be worn, the procedures to be observed following an accident or the names of people authorised to perform certain jobs. Although failure to abide by these regulations is not an offence which will lead to prosecution, such a failure may be regarded as a breach of the contract of employment and could lead to dismissal.

However, it is important to remember that evidence concerning any alleged breach of the law or other regulation would be required in order to prove whether the alleged breach did or did not occur, and that the severity of any penalty, ranging from a warning to imprisonment, would be related to the seriousness of the breach. This is covered in more detail later in the chapter.

4.2 THE HEALTH AND SAFETY AT WORK, ETC., ACT, 1974

This is undoubtedly the most important single piece of legislation in the United Kingdom to affect the working environment in recent years. Unlike previous safety legislation, this Act applies to all people at work (with the exception of domestic workers in private households) and it will simplify the state of the law; for example, it will supplant all or part of 31 other Acts and extend legal cover to approximately five million workers for the first time.

Additionally, the responsibility of workers towards the general public is spelled out for the first time.

Section 2 of the Act requires all employers to ensure, so far as is reasonably practicable, the health, safety and welfare at work of all their employees. 'Reasonably practicable', as it has been defined previously in case law, means that the cost of any measures may be related to the likely benefit, so that expensive precautions would not be required to eliminate slight risks. (If the Act had said 'as far as is practicable', any possible precautions would have to be taken, even though the cost was high and the risk slight; this would be similar to prohibiting tea-drinking because of the risk of drowning, and would obviously be absurd.) This duty is defined in the following ways

(1) the provision and maintenance of safe plant and equipment and safe systems of work

(2) arrangements for the safe use, handling, storage and transport of articles and substances

(3) the provision of information on matters of health and safety, and the provision of instruction, training and adequate supervision of employees

(4) the provision and maintenance of safe places of work and access thereto

(5) the provision of a safe and healthy working environment and also adequate welfare facilities and arrangements.

There is an additional duty placed on employers, by section 3(1), to conduct their undertakings in such a way as to ensure, as far as is reasonably practicable, that persons who are not in their employment are not exposed by their activities to health or safety risks.

The duties of employees are given in section 7, which requires them to take reasonable care for the safety of themselves and others at work, and to co-operate with their employers in carrying out statutory obligations. Section 8 prohibits any person from intentionally or recklessly interfering or mis-using anything required by statutory provision in the interests of safety, health and welfare.

Section 6 of the Act requires that those who design, manufacture, import or supply equipment or substances for use at work must ensure, so far as is reasonably practicable, that the equipment and substances shall be safe and without risk to health when properly used.

The Act also requires that employers must prepare, in writing, and publicise a statement of their safety policy. This will oblige employers to consider carefully how they intend to ensure a safe and healthy working environment, how they will give training and instruction and who has responsibility for carrying out the details of the policy.

The general duties mentioned above are quite new and go far beyond the requirements of previous legislation, and it is an offence to fail in any of these requirements. It is an offence, punishable by fine or imprisonment, to interfere with safety equipment; letting off fire extinguishers for fun or using carbon dioxide extinguishers instead of the proper cylinders as a source of solid carbon dioxide is just as illegal as removing lifebelts from a riverside.

The work place will have to be kept at a reasonable temperature and adequately ventilated (as was required by the Factories Acts and the Offices, Shops and Railway Premises Act) and employers who fail to do so will be in breach of the law — the employee's duty to co-operate could give legal support to complaints about cold or draughty laboratories! There will be many difficult situations when money for routine maintenance is insufficient to repair defects which have become dangerous. There is an obvious duty to report such defects, preferably in writing (keep a copy carefully); it seems unlikely that anyone who has behaved responsibly in such circumstances would find any legal difficulties, but it is now most unwise to dismiss reports from other people at work of situations that they consider to be dangerous. How often have complaints been greeted with such comments as 'Don't be so fussy, it seems perfectly safe to me. Besides we haven't got the money to repair it, so you will just have to carry on as best you can.' Such a reply might be given in evidence in a prosecution under the Health and Safety at Work Act. If a piece of equipment, or even a laboratory, is dangerous and for whatever reason cannot be made safe, there is no alternative but to cease to use it.

The duty of manufacturers to supply safe equipment will, of course, cover equipment made in laboratory workshops; be especially careful of second-hand equipment, because it may have been manufactured to lower safety standards than are now acceptable, and, if you were considered to be the supplier, the terms of section 6 could apply. Although the need to supply safe equipment is obvious enough, there can be problems. For example, some hi-fi equipment is not earthed because of the possibility of picking up unwanted 'hum', and at least one large education authority limits the use of portable power tools by pupils to tools supplied at 110 V, suggesting that they consider the normal 240 V too dangerous.

Many science technicians and teachers have been worried by their duty to take care for the safety of themselves and others; obviously they do not wish to have accidents, but the thought of being prosecuted after an incident when they had been as careful as they know how is very worrying. However, it is the clear duty of the employer to provide instruction, training and supervision, which would include the need to pass on new information concerning hazards. Therefore, provided the teacher or technician has complied with any instructions given to him, and has behaved responsibly, he should not find himself in any further trouble due to his own lack of knowledge or training. Of course, in many cases it may be difficult to know which person is regarded as the employer, since the superintendent, head of department, head of division, managing director or board of directors together with personnel department and company secretary all have some responsibilities in this area. However, as is explained later, the normal penalty will be either a notice to rectify a defect or a fine, both of which will affect the organisation as a whole, and imprisonment will only be imposed in serious cases, probably only in cases where an individual has been deliberately reckless or repeatedly at fault.

Section 2(4) allows 'recognised trade unions' to appoint 'safety representatives' and it is established in section 2(6) that the employer is required to consult these representatives on promoting and developing measures for health and safety, and in checking the effectiveness of such measures. The regulations made under this section allow for a safety committee on each site, comprised of at least one member appointed by each union with members on that site. Such a committee must be set up within three months of a request being made by at least two safety representatives. Obviously in a large industrial plant, there might be a dozen unions representing thousands of workers, whereas in a small primary school each of the six or so staff might be in a different union, meaning that the safety committee would comprise the whole staff. The safety representatives have the function to investigate potential hazards and examine the causes of accidents, investigate complaints relating to the health, safety or welfare of employees and make representations to the employer, meet with Health and Safety Inspectors, attend safety committee meetings and carry out inspections; for all these functions, the employer must allow the safety representatives to take the necessary time off work without loss of pay. The representative is also entitled to 'reasonable' training without loss of pay. Safety representatives have the right to inspect the premises at least once every 3 months, or following an accident, or where there has been a 'substantial change' in conditions of work. It must be emphasised, however, that these *functions* of safety representatives do not impose *duties* upon them; that is to say, they may carry out inspections, and the employer must allow them to do so, but if they do not, they cannot be penalised for their lack of enthusiasm or efficiency. On the other hand, the Act imposes many 'duties' on employers, and they can be penalised for failures, for example, in the provision of a safe place of work. The Safety Representatives and Safety Committee Regulations, 1977 are operative from 1 October 1978.

If the actions of an employee in the course of his employment lead to a claim for damages, the employer is responsible for paying these damages, usually through insurance. This principle, known as 'vicarious liability', has established that an employer is responsible for the effects of actions of his staff. However, the Health and Safety at Work Act has made an important change in the employee's situation, in that, in some circumstances, he may have committed an offence for which he may be prosecuted and fined or, in extreme cases, imprisoned if found guilty; the offence could arise, for example, from section 7, which requires reasonable care for others, or from section 8, which prohibits interference with safety equipment. Such actions as by-passing safety interlocks, failing to wear protective clothing or goggles or failing to clear up spillages are among other possible offences.

The responsibility for enforcing the provisions of the Act lies with the Health and Safety Inspectorate. These Inspectors have replaced the Inspectors previously authorised by the Factories Act, the Mines and Quarries Act, the Explosives Acts, etc. Their normal method of dealing with suspected breaches of the Health and Safety at Work Act will be to discuss the situation with the employer in the first instance. It seems likely that the majority of unsafe or unhealthy situations will be able to be dealt with in this way. The Inspector, however, has the authority to issue an Improvement Notice; this will state the contravention of the Act to which it refers, the reason why the Inspector considers an offence to have been committed and will require that the contravention shall be remedied within a specified period. The Inspector may also offer advice as to how the situation might be remedied. When an Improvement Notice has been issued it must be complied with within the specified period unless an appeal has been made to an Industrial Tribunal; failure to comply is an offence.

If there is a situation which the inspector considers involves a risk of serious personal injury, he may issue a Prohibition Notice, which may take effect immediately, or at any time that the Inspector decides. These notices demand that the activities specified in the notice must cease, either until such time as the matters complained of have been rectified, or until an Industrial Tribunal over-rules the notice. It is obvious that Improvement and Prohibition Notices give the Inspectorate substantial powers, and the sensible use of these procedures should bring about dramatic improvements in industrial safety.

Prosecutions may be instituted by the Inspectorate for breaches of the Act itself, as mentioned earlier, or for failure to comply with Improvement or Prohibition Notices, as well as for such offences as obstructing or impersonating an Inspector or making false entries in registers, etc. The normal penalty is a fine, although imprisonment may be ordered in serious cases. It seems most unlikely that any employee or employer who acts responsibly will incur any of the possible penalties, but the possibility of the managing director and maybe the board of directors being locked up if all else fails should be a powerful deterrent for the few rogues.

Eventually Regulations and Codes of Practice (with similar legal standing to the Highway Code) will be issued under the authority of the Health and Safety at Work Act. Until they are issued, the previous Regulations and Codes authorised by the Factories Act, Radioactive Substances Act, etc., will continue to apply.

Reasonable use of the provisions of the Health and Safety at Work Act will undoubtedly lead to great improvements in all aspects of safe working, although unreasonable use, by undue emphasis on minor details, or insistence on unreasonably expensive precautions, could have a seriously detrimental effect both on costs and the possibility of working without an excessively complex 'rule book'. Any individual worker should normally expect to raise matters concerning safety through his Safety Representative and Safety Committee. In extreme cases where a reasonable solution seems unattainable, the Health and Safety Inspectorate may be approached directly and normally they will be able to give advice and be able to use their position to help to achieve a solution.

4.3 DISPOSAL OF DANGEROUS SUBSTANCES

We have seen that the Health and Safety at Work Act requires employers to conduct their activities in such a way as not to expose the general public to risks to their health or safety. Obviously this must require the safe disposal of dangerous substances used in laboratories. Suppliers must also ensure that instructions concerning the proper use of their products are given.

Examples of the types of dangerous substances will include

(1) radioactive isotopes
(2) poisonous chemicals
(3) flammable substances
(4) pathogenic organisms
(5) corrosive materials
(6) materials likely to react with other substances to give hazardous products.

The Deposit of Poisonous Waste Act, 1972, prohibits the depositing of poisonous, noxious or polluting chemicals on land, but permits the tipping of such chemicals in certain places by authorised persons. Before such chemicals may be moved, the local authority and the water authority must be notified of the nature and chemical composition of the waste and the description of the containers, as well as where it will be removed from and where it will be going to, and who is responsible for undertaking the removal. There are several companies specialising in the disposal or recovery of chemical wastes: look under Waste Disposal Contractors in the Yellow Pages telephone directory. Advice may be obtained from the local authority, the scientific branch of the Greater London Council or the Hazardous Materials Service, Chemical Emergency Centre, Atomic Energy Research Establishment, Harwell, Didcot, Oxon.

The disposal of radioactive isotopes is covered by the provisions of the Code of Practice authorised by the Radioactive Substances Act, 1960. Normally the storage of radioactive waste is prohibited, and the quantities of activity that may be disposed of are specified. For open sources, small quantities in solution may be diluted and disposed of down the drains, and small quantities of solid, diluted with non-radioactive material, may be disposed of in the dustbin. The licence issued under the authority of the Code of Practice will give exact details.

Schools and colleges using only small quantities of radioactive material may prefer to work under the less onerous conditions of the Department of Education and Science Administrative Memorandum AM 2/76 which, with its explanatory notes, gives all the necessary information. Closed sources that are no longer required should be returned to the original supplier. Wastes that cannot be disposed of by these methods may be sent, by prior arrangement, to the Industrial Chemistry Group, Atomic Energy Research Establishment, Harwell. Of course, the only way in which radioactive isotopes can be finally disposed of is by storing them until their radioactivity has fallen to safe levels, although adequate dilution will render small quantities safe. Further advice can be obtained from the National Radiological Protection Board, Harwell. (A fee may be charged.)

Unlike radioactive isotopes, poisonous chemicals will remain poisonous for ever. In some cases, controlled tipping of waste may prove satisfactory, but normally a chemical treatment is necessary. This treatment may either break up the poisonous compounds or combine them in such a way that they become harmless. For example, sodium cyanide contains only sodium, carbon and nitrogen, all of which are essential to health.

Small quantities of many substances can be safely dealt with by generous dilution, but there are other substances for which this cannot be done safely. It is advisable to consult the original supplier if there is any doubt.

It is sometimes necessary to dispose of flammable liquids, for example, contaminated solvents or lubricants. If there is a regular supply of waste which cannot reasonably be reprocessed, it may be most satisfactorily disposed of by burning in a special incinerator, but smoke control regulations must be considered. If there is no special incinerator, small quantities may be evaporated in the open air or burnt, or the waste may be absorbed into cotton wool and then ignited in a suitable location well away from other fire risks. Volatile flammable liquids must not be disposed of down the drains — apart from contamination problems, the liquids will evaporate in the warmth of the sewers and form explosive pockets of vapour. Remember that many organic vapours are substantially more dense than air, and slow escapes of such gases may lead to the build-up of explosive mixtures with air in such places as below floorboards, in sealed sumps in storerooms or in laboratories with floors below ground level. It is essential to ensure that fume extraction systems blow the exhaust gases into the open air where they will be carried away safely.

There is a wide variety of possible pathogenic hazards which may be encountered. Information on the safe handling of these is given in chapter 7. It is important to remember that the general requirements of the Health and Safety at Work Act apply to the use of pathogens, even though they may not be visible.

The safe method of disposal of corrosive materials is by neutralisation. Thus acids may be reacted with a convenient, cheap basic material until the product is neutral. It is important that this process should be carried out conscientiously; a lorry driver was killed when he tipped an acid waste on the wrong part of a dump. Unfortunately there were sulphides present, and the driver was overcome by the resulting hydrogen sulphide, a gas which has a strong smell in small quantities, but cannot be smelt in large concentrations because of its effect on the nerves in the nose.

Many other substances require careful disposal. Nitrates will give off nitrogen dioxide from the reaction with sulphuric acid, and calcium carbide reacts with water to give acetylene gas. There are reference books which list the hazards that arise from mixtures of chemicals; although these are expensive, copies should be available for consultation when required.

It should be normal practice for local regulations to be issued by employers stating requirements for the safe storage and disposal of hazardous materials. The Petroleum Acts, for example, have established detailed regulations for the safe storage of highly flammable liquids and liquefied petroleum gases. The Radioactive Substances Act has authorised the Code of Practice, which demands that adequate arrangements shall be made, but leaves most of the detail to be worked out by the locally appointed Radiological Protection Officer. The use of radioactive isotopes, aromatic amines, mercury and certain biological organisms, is controlled in schools by regulations issued by the Department of Education and Science.

It is unfortunately true that there is no simple source of information on the hazards of specific materials. The long delays between diagnoses of asbestosis and mesothelioma and the issue of the Asbestos Regulations illustrate the difficulties that arise in accurately assessing the nature of hazards, and the emotions that have surrounded the discussions of links between smoking and lung cancer show the problems in getting the level of hazard understood. There are several very good reference books which explain the hazards and correct procedures to be followed with many dangerous substances; some of these books are listed at the end of this volume.

4.4 TRADING ACTIVITIES

The law relating to the many aspects of consumer protection has led to several important changes in the law concerning laboratory trading activities, and doubtless there will be more changes to come.

The Sale of Goods Act, 1893 requires that goods sold should correspond to their description, be of 'merchantable quality', and be fit for the purpose for which they were sold. The duty of supplying satisfactory goods lies with the retailer, or the seller from whom the purchase was made, since this is the person with whom the contract to buy was made. Any 'guarantee' offered by the manufacturer can only give extra protection and, by recent legislation [Supply of Goods (Implied Terms) Act 1973], may not take away rights which would have otherwise existed. (It should be noted that this latter Act only applies to private consumer sales, and does not cover business sales to institutions such as laboratories.) The Trade Descriptions Act, 1968 tightened up the conditions applying to descriptions and statements made concerning goods on sale. These Acts, together with the provisions under the Health and Safety at Work Act to supply safe equipment and substances, make it illegal to sell unsatisfactory or dangerous goods. Of course, it is essential to remember that the buyer must act responsibly; an article of low but acceptable quality at a low price cannot be expected to be as good as a better quality article at a higher price, and no legal protection can apply to cases where goods are used for a purpose which the supplier could not easily have foreseen. The normal provisions of sueing in the courts for damages, in the form of monetary compensation for losses incurred due to unsatisfactory goods, are available, but the cost of taking a supplier to court may be substantial and the risks of losing the case could mean that legal action of this type was not worth starting. If normal attempts to obtain satisfaction by direct approaches to the suppliers are unsatisfactory, the Trading Standards Department of the local authority (formerly the Weights and Measures Department) should be consulted. This Department has the duty of ensuring the enforcement of the provisions of most legislation relating to sales.

It must be noted that, at present, the law relating to the provision of services, for example, maintenance or advice, is far behind the law relating to the provision of goods in the protection that is given. There will be little possibility of compensation if an adviser acts in good faith but his advice turns out to be wrong, unless he can be shown to have been negligent. This contrasts with the position of a retailer who has responsibilities even when he sells goods which are defective in spite of his best endeavours to ensure their quality (although in this type of situation, the retailer would not be guilty of a criminal offence).

4.4.1 Purchase of Alcohol

The Customs and Excise Act, 1952 covers the purchase, storage and use of alcohols, and makes the private manufacture and distillation of alcohols illegal. However, applications for the duty-free purchase of alcohol may be made, provided that the alcohol will be used for authorised purposes, which include educational and scientific laboratory usage. An application must be made through the local Customs and Excise office, whose address is in the telephone directory, stating the purpose for which the alcohol is required, the premises for its storage and the approximate annual requirement. A requisition book will be issued if the application is approved, and a requisition must accompany each order sent to the supplier. When the alcohol is received, it is accompanied by an official permit which must be given to the local Excise Officer. A stock book, available for inspection by the Excise Officer, must show all receipts and issues of alcohol and an annual return on the appropriate form must also be made. Duty-free alcohol must be kept under lock and key separate from duty-paid spirits and methylated spirits.

4.4.2 Purchase of Poisons

Poisonous substances should obviously be treated with care, and safely stored when not in use. The Poisons Act, 1972 is mainly concerned with the retail sale of poisons. Part I poisons may only be sold by a retail pharmacy, but Part II poisons may also be sold by persons whose name is on the local authority list, for example, weedkillers may be sold by a garden centre. Part I poisons may be sold only to people having a certificate properly authorised, for example, by certain police officers, or to people 'known to the seller as persons to whom the substance may be properly sold', and a record of the transaction must be kept and signed by the purchaser (the Poisons Book). The Act makes labelling requirements, that is, the label must bear the name of the substance, the name and address of the supplier and the word POISON; it also gives guidelines on how to classify substances into either Part I or Part II and allows sales to wholesalers, doctors, hospitals, etc., to scientific research or educational institutions or for export.

Further advice may be obtained from the local police station.

4.4.3 Purchase of Goods from Abroad

Goods imported from outside the European Economic Community (Common Market) area are subject to taxation, which is intended to protect home industries. The Import Duties Act, 1958 and subsequent Finance Acts have included the conditions under which this import duty taxation may be excused. The important condition is that there must be no Common Market equivalent; this normally means that the nearest available equipment must fail to be satisfactory in some important respect, but the fact that an adequate alternative is more expensive is usually not a satisfactory reason. The imported article must not be for sale or for use for any commercial purpose, but must be required for use in scientific research or education. It is essential to give notice to the Department of Trade of intention to apply for relief from duty when the order is placed, and this relief must be granted before the goods are paid for — any duty that has already been paid cannot be reclaimed. Further details are available from the Duty Remission Branch of the Department of Trade.

4.5 EXPERIMENTS WITH ANIMALS

The Cruelty to Animals Act, 1876 forbids experiments on animals which will cause pain. It is not necessary to consider details here, but there are precise conditions laid down to govern the circumstances in which experiments, normally under anaesthetic, may be undertaken. A Home Office licence is required, which may only be issued to a named person working in named premises. The Act does not cover experiments which do not cause physical pain, for example, learning experiments are permitted. Animals required for dissections, etc., must be humanely and painlessly killed.

4.6 ASSIGNMENTS

4.1 What particular Statutes or Regulations apply to the activities that you undertake (a) at work, (b) at college? Are copies available to you?

4.2 Are there any features of your place of work that seem to you to be unsafe? By considering costs, would it be 'reasonably practicable to make the changes necessary to improve safety?

4.3 Have you read your employer's statement of safety policy which is required by the Health and Safety at Work Act?

4.4 What Codes of Practice apply at your place of work? In which year were they issued, and by whom?

5 Safety in Chemistry Laboratories

5.1 GENERAL PRECAUTIONS

Chemistry laboratories can be the most dangerous places, but if one is aware of the dangers and potential hazards that can and do exist, the laboratory will be no more dangerous than the average work place.

The key to laboratory safety is awareness and careful working, that is, common sense. Be aware of the potential dangers inherent in a chemical procedure and take appropriate precautions, no matter how trivial they may seem, to ensure that an accident does not occur. Not many years ago a young school girl attempted to grind some crystals of potassium chlorate in a mortar; the resultant explosion blinded her for life. In another laboratory, a girl who was heating iron filings with sulphur received the red hot mixture in her face because she had not been warned that the reaction generated heat and would become violent if she continued to apply external heat. In neither of these cases were goggles or face masks being worn, and the people concerned probably did not realise that there was any danger. Do not be afraid to ask 'what precautions must I take in order to do this job safely?' The price of ignorance may be too high to pay. If your fellow workers can't give you a satisfactory answer, seek advice from a suitable reference book (see the list at the end of this book). Always be aware that others are working near you and take care that whatever you may do will not harm them if something unforeseen occurs. Above all, treat all chemicals with respect; do not inhale fumes (except very cautiously for identification purposes), do not taste any chemical material and do not handle any chemical with bare hands. Eating and drinking in the laboratory must be prohibited, and because of the obvious danger of fire, smoking must also be forbidden. Scrupulous attention must be paid to personal hygiene; on leaving the laboratory, wash your hands thoroughly and remove your laboratory coat to avoid the risk of taking contamination elsewhere. Failure to be careful could lead to toxic materials being ingested or perhaps lead to the development of dermatitis, a common, irritating and persistent skin disease among some chemists.

Minor accidents are bound to happen, even in the best regulated laboratories; a beaker of hot liquid may crack, spilling its contents over the bench and worker; another worker may not have taken sufficient care, and the contents of a test tube may be splattered over you. For these reasons a laboratory coat must always be worn; if it becomes soaked in a hot and corrosive liquid it can be removed quickly and easily, and so provide a fair degree of protection for clothing and person. Resist the temptation to roll up the sleeves of the coat, even in hot weather, for this immediately exposes the forearms to damage. Keep the front of the coat buttoned up; the medical type of coat which buttons to the neck and has a generous overlap will prevent material getting inside and will give a double thickness at the front. Cotton drill coats are quite absorbent and, if they catch fire, will fall apart; nylon and similar synthetics will not absorb liquids so well and give less protection and, if they burn, will form into hot sticky globules which can cause very serious skin burns. For certain tasks more extensive personal protection will be required; for example, where there is a danger of a liquid spitting (as in the dilution of sulphuric acid) or irritating gases being evolved (as in the manipulation of bromine) the eyes must be protected by using goggles or a face visor. It is good practice to wear safety glasses at all times and in some laboratories it is obligatory to do so. Prescription spectacles can be made of safety glass and more enlightened employers have provided them, when necessary, for many years. If any chemical should enter the eye, then wash the eye copiously with water — don't wait around for the correct eyewash to be found, just put the head under the nearest tap and make sure that the eye is well washed, then seek medical advice. It would seem appropriate here to mention the use of rubber gloves as an item of personal protection. There are many cases where the use of gloves is essential to protect the hands from contamination, for example, in the handling of radioisotopes, poisons or carcinogens but there is a tendency, especially among women, to wear gloves almost permanently when carrying out chemical operations to 'protect their skin'. In many cases it will have the opposite effect in that dermatitis can develop on hands permanently soaked in sweat and also there are very few gloves available with sufficient grip to cope with slippery glassware. Finally, some laboratories use materials that, if spilt, would require the use of a respirator to deal with the spillage. It is important that you know how to use the respirator

properly — don't wait for the spillage to occur before finding out how to use it.

5.2 FUME CUPBOARDS

Although there are many different types of fume cupboard available today they are all of the same basic design and have the same function, that is, to provide a totally enclosed, well-ventilated environment in which a chemical reaction can be safely carried out, so that any fumes that develop can be removed efficiently and, if the reaction should explode, the explosion and products (including broken glassware) can be contained. At the

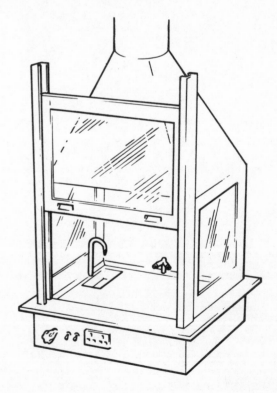

Figure 5.1

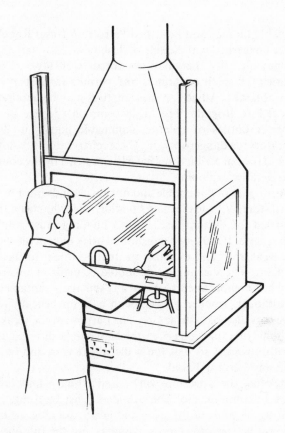

Figure 5.2

reaction has ceased and that any noxious or flammable vapours have been removed by the ventilating system. The fume cupboard is essential for the manipulation of large quantities of flammable liquids or liquids which give off dangerous or irritating fumes, for example, all reactions involving the use of ether (such as a Grignard reaction), liquid ammonia, sulphur dioxide or carbon disulphide, must be carried out in a fume cupboard.

It is important that the rate of air flow into the fume cupboard is adequate to prevent the fumes coming back into the laboratory. It is recommended by the Department of Education and Science for school-level work that the air flow when the sash is raised 500 mm should be at least 0.3 to 0.5 metres per second depending on the degree of hazard involved. The speed should be higher as the hazard increases. The flow should be checked at several places in the plane of the opening, the lowest value being the significant one. The direction and uniformity of the air flow can be checked using either a candle or smoke pellets. If a satisfactory air flow cannot be obtained with a 500 mm opening, then the glass should be lowered until the flow is adequate and the new position clearly marked.

To avoid the expense of installing ducting in an existing building, and also to reduce the heat loss due to warm air flowing from the building to waste, recirculating fume cupboards are becoming available. These use a replaceable micro-filter which is fine enough to remove the large molecules of the pollutants but allows the smaller air molecules to pass through. These cupboards require extra care due to the obvious dangers of recirculating fumes should the filter system fail.

It is unfortunate that in many laboratories, fume cupboards are also used to store 'temporarily' unwanted reaction products or bottles of dangerous chemicals; in fact the cupboard can quickly become the repository of a collection of bottles of highly dangerous materials which will soon lose whatever labels they may have had. It is important, therefore, that all fume cupboards should be cleared out and cleaned at regular intervals, perhaps fortnightly. Remember that the fume cupboard is to work in; it is not a short or long-term store.

same time, free access to the reaction vessel is necessary and a supply of the usual services, for example, gas, water and electricity, is required. A common type of fume cupboard is shown in figure 5.1 and consists of a wooden frame filled with safety glass, a sliding door and exhaust vent. The door may be raised completly to allow equipment to be set up, or slightly so as to allow the hands to enter and yet protect the major part of the body (figure 5.2).

The controls for all the services must be on the outside and the base of the cupboard 'dished' so as to contain any liquid that might be spilt. Should a reaction get out of control, the front of the cupboard can quickly be closed until it is certain that all further

5.3 FLAMMABLE LIQUIDS

Fire is by far the most common of laboratory hazards, brought about by the indiscriminate use of flammable liquids near a naked flame or hot surface, or by the careless disposal of flammable solvents down the sink where they may react with other materials that have been inadequately washed away. The only flammable solvents that may be disposed of down the sink are those which are miscible with water in all proportions, such as alcohol, acetone and acetic acid which do not form pockets of explosive vapour in the sewers. All other solvents must be disposed of by burning or burying, unless recovery is possible.

For general purposes, flammable solvents may be divided into two broad categories: (I) those of low flash or ignition point and (II) those of high flash point. Solvents of low flash point such as ether, 40/60 petroleum ether, carbon disulphide and acetone must always be used in the fume cupboard. All of these solvents have a very low flash point (below 0 °C) and heavy vapours—it is not unknown for these vapours to 'crawl' along the laboratory floor and be ignited by a flame at the other end of the laboratory. (See section 3.1.)

Many chemical operations require the use of dry (water-free) solvents such as ether or benzene; these solvents are commonly rendered water free by squeezing sodium wire into the bottle of solvent. It is of the utmost importance that when all the solvent has been used, the sodium residue be made harmless. This can be done simply and safely by pouring some methanol on to the sodium residue, the sodium dissolving in the methanol and the resultant sodium methoxide being washed out with water. Many operators in chemical factory bottle-washing plants have been badly injured because sodium has been left in discarded ether bottles.

Fires in laboratories, although fairly common, are usually easily and quickly dealt with and rarely become dangerous unless large quantities of flammable liquids are held in the laboratory. For this reason, no laboratory should contain more flammable solvents than are needed for day-to-day running, the bulk supply being held elsewhere. The Petroleum (Consolidation) Act, 1928 limits the storage of liquids with a flash point below 23 °C to a maximum of 13 litres unless a permit for more has been granted. The Highly Flammable Liquids and Liquefied Petroleum Gases Regulations, 1972 are concerned with liquids of flash point less than 32 °C; the maximum quantity that may be stored is 50 litres. There are exceptions for small containers and permits for larger volumes may be obtained. Advice on these matters may be obtained from the local Fire Brigade. The fairly common practice of storing Winchester bottles of volatile, flammable liquids in domestic refrigerators is a dangerous one. These refrigerators are not flame-proof and several serious fires have been caused by the contents of the refrigerator exploding.

Bulk supplies of flammable and noxious liquids must be kept in a special store, outside the main building, which has been specially constructed for the purpose. Such a building must conform to certain criteria. It must be a substantial brick or concrete building, with a weak roof and excellent ventilation direct to the outside atmosphere. The shelving inside should be made of concrete and divided into compartments, so that any spillage is contained by the shelf, although wooden shelves with a sump beneath to catch spillage may also be satisfactory. The purpose of the weak roof is that, should an explosion occur, the blast will be directed upwards and not outwards, so that the walls will remain and be able to contain any burning liquid.

Quite often, the same store will be used to hold cylinders of gases that are in common use. The cylinders must be firmly chained upright to the walls of the store and great care must be taken to distinguish between the empty cylinders and the full ones.

Laboratory solvents and liquid chemicals are usually purchased and stored in large glass bottles containing 2½ litres (the Winchester quart). The transport of these large glass bottles can present some hazard in that it is not unknown for the bottle to crack, either around the base or around the neck, particularly if the bottle has been stored in a cold place. Winchesters must always be carried in a suitable bottle carrier (figure 5.3), which will contain the contents in the event of a breakage, and *never* by holding them by the neck or hugging them to the body. Little imagination is needed to foresee the result of a bottle of concentrated sulphuric acid breaking in transit, especially if it is not in a bottle holder.

The same care must be applied to moving gas cylinders from one place to another. Always use a cylinder trolley, never attempt to roll the cylinder along the ground or roll it while holding the valve. In no circumstances must a cylinder of gas be left in a laboratory connected to apparatus and supplying gas unless it is under supervision. There is an incident recorded of a carbon dioxide cylinder pressurised at 900 p.s.i. (6×10^6 Pa) which lost its valve due to mishandling. The cylinder travelled like a rocket twenty feet across the floor, knocking a painter from a seven-foot scaffold, then turned round, bounced from a wall, chased an electrician down a forty-foot long room, again bounced off a wall and after going through a doorway went another sixty feet down a corridor before eventually falling into a truck well. Oxygen and nitrogen cylinders are pressurised at over 2000 p.s.i. (approximately 2×10^7 Pa)—the enormous damage that could result from carelessness may be imagined.

Care must be taken with the identification and use of gases from cylinders. BS 349:1973 Identification of the Contents of Industrial Gas Containers recommends the principles to be used in the marking of cylinders. This is done primarily by labelling the name and formula on the shoulder of the cylinder, and secondarily by a colour code. There are about thirty standard gases each having its own code; table 5.1 gives some examples.

Table 5.1

Gas	Colour of Cylinder	Colour of Band on Shoulder of Cylinder
Acetylene	Maroon 541	—
Ammonia	Black	Signal red and golden yellow
Argon	Peacock blue 103	—
Carbon dioxide	Black	French grey 630
Carbon monoxide	Signal red 537	Golden yellow 356
Helium	Middle brown 411	—
Hydrogen	Signal red 537	—
Neon	Middle brown 411	Black
Oxygen	Black	—

The colour of the band on the shoulder denotes the hazards associated with the contents, signal red 537 indicating flammable gases, golden yellow 356 indicating toxic gases and red above yellow indicating both hazards. The exact colours and their names and numbers are as defined in BS 381C. Full colour cylinder

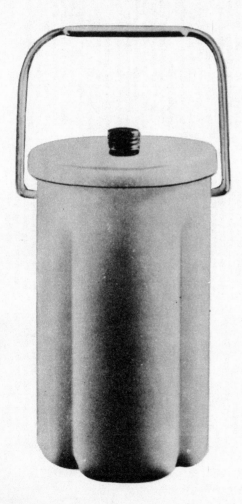

Figure 5.3 Winchester bottle carrier

Air Products Limited
Gas data and safety sheet 56
Hydrogen sulphide H₂S

Air Products

Specification

(Check current catalogue)

CP grade
Purity 99.6%

Typical analysis

By weight (liquid phase)
Carbon dioxide	0.13%
Carbon disulphide	0.09%
Sulphur dioxide	0.05%
Methyl mercaptan	0.02%
Carbonyl sulphide	0.01%
Methane	trace
Balance	air

Physical properties

Molecular weight	34.08
Vapour pressure (20°C)	17.38 Bar (g)
Specific volume (20°C, 1 atm)	701 ml/g
Boiling point (1 atm)	—60.33 C
Density, gas (0°C, 1 atm)	1.539 g/l
Density, liquid (B.pt)	0.993 g/ml
Critical temperature	100.4 C
Flammable limits in air	4.3—45% (by vol)
Specific gravity, gas (Air=1)	1.189

Colourless
Offensive odour of rotting eggs if above 1 vpm

Safety

WARNING: TOXIC, CORROSIVE, FLAMMABLE SUBSTANCE

Hazardous properties

Toxic tlv 10 vpm.
Slightly heavier than air.
Liquid in cylinder under gas at 17 Bar (g) pressure.
Liquid and gas corrosive to skin and eyes.
Liquid on skin causes cold burns.

Safety precautions

(see storage and handling instructions on reverse side)
Store cylinders outside in open air and away from oxidants.
Cylinders in use should be in open air or in a force ventilated fume room.
Wear protective clothing (rubber gloves and apron, chemical safety goggles).
Earth all equipment.
Connect via suck back trap.
Personnel should wear H₂S monitors.
Keep self-contained positive pressure breathing apparatus nearby.

Material compatibility

Corrosive when moist.
Stress cracking may be caused by moist hydrogen sulphide.
Aluminium is preferred. For dry hydrogen sulphide iron or steel are satisfactory. Brass, though tarnished, is also acceptable.

All equipment must be adequately designed to withstand the process pressures to be encountered.

Leak detection

Put on positive pressure breathing apparatus. Apply soap solution to suspect sites in lines and equipment, bubbling shows up leaks. Use detector tube selective for hydrogen sulphide.

Toxicity

The odour although distinctive is not always a satisfactory warning of exposure. Inhalation can cause depression of the central nervous system, sometimes stimulation and ultimately death by respiratory paralysis. 100 vpm produces slight symptoms after several hours exposure. 200 vpm can be tolerated for 1 hour without serious consequences. 500 vpm is dangerous within 1 hour.

Acute poisoning (over 700 vpm) is the greatest danger causing hyperpnea leading to death by respiratory paralysis. At high concentrations unconsciousness and collapse can occur within seconds, when it can cause inflammation of the respiratory tract and pulmonary oedema, accompanied by headache pains in the chest and difficulty with breathing. At above 50 vpm there may be eye irritation and conjunctivitis.

Symptoms

For subacute poisoning they include headache, dizziness, excitement, nausea, dryness and pain in respiratory tract, and coughing.

EMERGENCY ACTION

In the event of an accident or emergency the instructions below must be implemented immediately. After emergency action has been taken, contact Air Products Limited at Crewe (0270) 583131 (24 hour service) for further instructions.

Inhalation

Minimising personal risk, immediately remove victim to uncontaminated area. Ensure there is no obstruction to the airway. Administer pure oxygen. If breathing weak or stopped apply artificial respiration with simultaneous administration of oxygen, preferably using oxygen resuscitator. Summon ambulance. Keep warm and rested.

Eye or skin contact

Thoroughly irrigate affected area with water for at least 15 minutes. Seek medical advice as soon as possible.

Leaking cylinder

If cylinder in enclosed area, evacuate the area. Put on positive breathing apparatus. Check cylinder valve closed. Move cylinder to fume room or open space downwind and away from persons. Post warning notices and seal off area.

Action in the event of fire

Put on positive pressure breathing apparatus. Quick action to close valve may extinguish fire from valve outlet (care!). In general vacate area, call fire brigade. If unable to extinguish fire, keep surrounding cylinders cool with water hosed from a safe distance. Inform fire brigade of potential danger of exploding and rocketing cylinders.

Figure 5.4 Air Products Ltd gas data and safety sheet for hydrogen sulphide, H₂S

identification charts are obtainable from the suppliers of industrial gases.

Great care must be taken when using compressed gases from cylinders. Cylinders containing combustible gases have left-hand threads while those containing non-combustible gases have right-hand threads to prevent the interchange of fittings. Detailed advice on the safe handling and use of compressed gases may be obtained from the suppliers; for example, the pamphlets *Safety in the Use of Compressed Gas Cylinders* and *Handbook of Operating Instructions for Gas Welding and Cutting* available from the British Oxygen Company contain much useful information.

5.4 DANGEROUS CHEMICALS

The preceding section has been devoted to chemicals which are dangerous due to their flammability. However, there are many chemicals which are dangerous due to their toxicity or long-term physiological effects.

In certain cases, such as materials classed as dangerous drugs, the law imposes vigorous conditions as to the storage and use of such materials, but in other cases no specific legal obligation exists; however, this does not mean that no control should be exercised. It is important to remember that the general obligations of the Health and Safety at Work Act 1974 must always be observed.

Substances classed as dangerous drugs, for example, morphine, cocaine, phenobarbiturates, must always be kept in a secure locked cupboard. A careful and accurate record must be kept of the purchases made, the stock held, the amounts issued, to whom the issue was made, and for what purpose. These records, as well as the stock itself, are subject to inspection by Home Office Inspectors. Similarly strict control must be exercised over radioactive material.

One other chemical that is subject to strict control, in that it must be kept locked away and an accurate record kept of its use, is duty-free alcohol. This is not because of its toxicity, but simply to ensure that it is used for the purpose for which the licence was issued permitting its purchase without the payment of excise duty. Further details are given in chapter 4.

However, there are many chemicals which are quite free of legal obligation to control or even keep locked away; but it is in the interests of general safety and of the person responsible for the laboratory to exercise strict control over their storage and to ensure that they are issued only to responsible persons. It is very difficult to draw up a specific list of such materials and only a general guideline can be laid down. It is suggested that all materials which

(1) can cause cancer by absorption, ingestion or inhalation over a period of time, that is, carcinogens

(2) are virulent poisons, such as inorganic cyanides or organo-phosphorus poisons

(3) are cumulative poisons, for example, mercury

(4) exhibit violent chemical activity, such as alkali metals and phosphorus

be subject to strict control in their storage and use. For example, a short list could be

inorganic cyanides	arsenical compounds
oxalic acid and its salts	barium and its salts
mercury and its salts	lead salts
organo-phosphorus compounds	yellow phosphorus
alkali metals	alkyl and aryl nitrosamines

Special care must be taken when inorganic cyanides are used; these must never be issued without the appropriate antidote (see note at the end of the book) because their action is very rapid. Many substances of biological and chemical origin are capable of causing changes in body proteins as a result of exposure to minute doses of the order of nanograms. These small sensitising doses may have no apparent effect but subsequent exposure to even smaller doses of the same or related chemicals can cause startlingly rapid, very severe, reactions. Examples of these substances, known as sensitisers, are 1-bromo-2, 4-dinitrobenzene and the similar chloro- and fluoro-compounds. Sensitisation can occur as a result of skin contact, ingestion or inhalation; great care should be taken in the use of all unfamiliar substances, especially biochemicals.

It is difficult to give a comprehensive list of poisons and carcinogens here, but there is a wealth of literature on the subject

Oxidising substance   **Corrosive substance**

Causes severe burns. May cause skin ulceration. Powerful
oxidising agent. Contact with combustible material may
cause fire. Wash immediately after using or in case of skin
contact. If eyes affected wash with water, if any swallowed
wash out mouth, drink water followed by milk of magnesia.
Get medical aid in both cases and also if you feel unwell.

EDUCHEM®

FOR EDUCATION

318129
Chromium(VI) oxide
(Chromium trioxide)

CrO₃ RMM 99·99
RECRYSTALLISED

SPECIFICATION
Assay 98·0% min.
Maximum Limits of Impurities

	per cent
Sulphate (SO₄)	0·06
Iron (Fe)	0·01
Sodium (Na)	0·05

HW HOPKIN & WILLIAMS
CHADWELL HEATH ESSEX · ENGLAND

A **SEARLE** COMPANY

PEEL BACK

CCl₄

TETRACHLOROMETHANE
CARBON TETRACHLORIDE

M,154

WARNING

Harmful vapour—causes liver damage.
Avoid breathing vapour.
When heated to decomposition gives off phosgene.
Wear respirator when handling large quantities.
Do not smoke.

EMERGENCY ACTION

Eyes —Irrigate with water.
Lungs —Remove from area and keep at rest. If severe
        get medical attention.
Mouth —Wash out with water.
Skin —Drench with water.
Spillage—Add detergent and drench with water.

ASSOCIATION FOR SCIENCE EDUCATION

Figure 5.5

that can be consulted (see the list at the end of this book), particularly G. D. Muir, *Hazards in the Chemical Laboratory* published by the Royal Institute of Chemistry which is a source of information on the toxicity and hazards associated with many chemicals and the treatment of poisoning by them.

5.5 LABELLING

The need for adequate labelling of chemical containers is quite obvious, but what information should the label carry? It is necessary here to draw a distinction between the labels affixed by manufacturers to bottles of their products, and the labelling of bottles of solutions and reagents prepared in the laboratory.

The requirements for labels on commercially supplied chemicals are quite clear, and, since the entry of the United Kingdom into the Common Market, should follow Common Market directives. In short, each label must carry the name of the chemical in large letters, an appropriate danger symbol, a risk phrase, and optionally, a safety advice phrase. If the material is subject to the Poisons and Pharmacy Act, 1933, or subsequent related legislation, it must also bear the legend POISON in red capital letters. Examples of labels are shown in figure 5.5.

In this definition of adequate labelling, two points need some explanation: first the nomenclature used to describe the chemical and secondly the nature of the risk phrase. Although the IUPAC system of nomenclature has been advocated and in fairly general use for many years, trivial names still linger, and are likely to do so for many years to come. In such cases it is customary to print the trivial name as well as the IUPAC name. This trivial name is usually printed underneath the systematic name thus

PROPAN-2-OL
(isopropyl alcohol)

The risk associated with the chemical must also be indicated by the use of a suitable phrase, which of necessity indicates the principal hazard only, for example,

FLAMMABLE
IRRITATING TO SKIN AND EYES

or

CONTACT WITH COMBUSTIBLE MATERIAL
MAY CAUSE FIRE
HARMFUL IF TAKEN INTERNALLY

The risk phrase may be followed by a safety advice phrase, provided the label is large enough, such as 'prevent contact with the skin' or 'avoid contact with skin and eyes'. Note that the terms 'prevent' and 'avoid' have quite different meanings. Prevent is a distinct injunction to use appropriate protective clothing, whereas avoid means that whereas due care must be exercised in the handling of the chemical, no special precautions are essential to one's safety.

The label, therefore, bears quite a lot of useful information, but if one is handling a material for the first time it is well worth while looking up further information on its hazards and treatment of spillage in the literature before using the material.

Laboratories always contain bottles of reagents or chemicals, either prepared in the laboratory or simply filled from larger bulk supply; these too must be adequately labelled. For many years small reagent bottles with etched labels have been available; in many cases these are quite adequate, but since they bear no indication of the hazard associated with the reagent, they should only be used for common well-known materials. The Association for Science Education publish a series of about 100 labels for use on reagent bottles which are based on Common Market requirements, and the Hopkin and Williams Educhem label series is a good example of a commercial system. However, it is often necessary to prepare bottles of solution for immediate use and in this case it is sufficient to label the bottle with the name of the solution, its concentration and the date of preparation. If special care has to be taken, a brief warning of the hazard should be included. Take care to use indelible ink for the label, be sure that the label is accurate, and is removed from the bottle once the contents have been used. To avoid any risk of ingestion of harmful material, self-adhesive labels should always be used in preference to those that tempt people to lick them.

Finally it is worth noting that the prolonged storage of chemicals, particularly in outdoor storage, often leads to labels falling off their containers. This can be prevented by coating the label with a thin film of wax (the paraffin wax is melted in a tin, and a thin coat brushed over the label and for about 2 cm all around it).

5.6 ASSIGNMENTS

5.1 Is there any instruction, either at work or at college, concerning the maximum amount a fume cupboard may be opened while still giving an adequate through draught?

5.2 What procedure is operated at work for the control and registration of poisonous substances?

5.3 What reference materials are readily available to you in which you can look up the hazards you might expect to be associated with a given chemical?

5.4 What rules are in force at your place of work concerning the transportation of large bottles of chemicals, for example, Winchester bottles?

5.5 If you make up a solution which is to be kept for two weeks in a glass stoppered bottle, what is the minimum information you should include on the label?

5.6 Are there any known carcinogens handled in the laboratory where you work? If so, what precautions are taken to guard against this hazard?

6 Safety in Physics Laboratories

6.1 GENERAL PRECAUTIONS

Physics laboratories are often considered to be less dangerous than some others such as chemistry laboratories. It is true that the hazards are less obvious but none the less real and ever present. As in most laboratories many risks can be minimised by taking care and giving thought to the possible dangers inherent in any specific situation. It is important always to remember that although one individual who has set up a particular experimental situation may be aware of the dangers associated with it, for example, exposed HT terminals or a laser beam, this individual may be called away at any time leaving what amounts to a booby trap for the next worker who comes into that laboratory. The only safe way to work in a laboratory is to bear in mind, at all times, the possibility of having to leave the room and another worker, or in some cases, member of the public, coming into that room — is that person at risk? If the answer is yes then your experimental arrangement must be modified to make it safe.

There are many experiments that require equipment to be left running overnight or during holiday periods; it may be that laboratories are locked during these periods and so, since people are excluded, it may appear safe to leave situations that could not be left unattended during the day. This may not be true. In some laboratory blocks it is usual for cleaners to come round in the early morning before work starts. These people are rarely seen, so it is easy to forget about them and the fact that they will have a pass key to every room, and their duties may well include at least cleaning the floor in the laboratories.

Whenever equipment is left running at such times, it is important to allow for the unexpected happening. A water-cooled diffusion pump was left running overnight, but at some time that night one of the water pipes detached itself from the pump. The tap was still turned on and the resulting flood caused expensive damage to the flooring and ruined some electrical equipment nearby. It was admitted that the ends of the pipe should have been secured in some way and all connections were then secured with jubilee clips. A week later, one of the rubber pipes split with more or less the same consequences and only then was it remembered that such water connections should be made with reinforced tubing.

Where a flow of gas is to be maintained for some length of time precautions should be taken to prevent possible sources of leakage such as a tube becoming detached. Rubber tubing should be routed so that it avoids hot parts of the apparatus and parts that could become hot if, for example, a flow of cooling water were interrupted.

Overheating is a common cause of difficulty in electrically powered equipment left running for some time. Several firms manufacture temperature-sensitive switches which are relatively inexpensive and can be used to switch off the supply in the event of the temperature of a piece of equipment rising above preset limits. Similar switches are available which sense the flow of cooling water; if this is reduced below a specified amount the electricity supply to the apparatus is switched off.

6.2 USE OF IONISING RADIATIONS

Radioactive sources are of considerable use to research and the general technology based on nuclear energy is of great benefit to mankind, but the radiations emitted by radioactive material can be extremely harmful, especially since our five major senses are not sensitive to them. These radiations cannot be seen, smelt or touched, and the effects of overexposure may not become apparent for several days, weeks or even years after an exposure incident.

This doesn't mean that sources of radiation should not be used — the benefits are too great — but they must be used carefully and with proper precautions (see figure 6.1) so as to reduce any one person's exposure to radiation to the absolute minimum possible. Nobody would suggest that children play on a motorway but, used sensibly, these roads are extremely useful and perfectly safe.

A radioactive substance is one where the nuclei of some of the constituent atoms are unstable, that is, they are liable, at some time in the future, to break up into smaller pieces or lighter atoms and in this process will emit one or more types of radiation. It is these radiations that are potentially harmful.

Two units are commonly used to quantify discussion on radiation. The first is a measure of how much radiation is being

Figure 6.1   Symbol to warn of ionising radiation hazard

emitted by a radioactive substance every second, called the becquerel (abbreviated Bq) and is the number of atoms breaking into smaller parts every second. (Until recently the unit used to measure the amount of radiation was the curie, abbreviated Ci. $1 \, Ci = 3.7 \times 10^{10} \, Bq$. The becquerel has been introduced to bring radiation units into line with the SI system.) The becquerel is a relatively small unit; one of the small demonstration sources that might be used in a secondary school would be of the order of $1 \times 10^5 \, Bq \, (2.7 \, \mu Ci)$ whereas a source such as might be used for industrial radiography might have as high an activity as $10^{11} \, Bq$ $(2.7 \, Ci)$.

The radiations that are emitted by these unstable nuclei may be absorbed by the body and the energy of the radiation thus transmitted to the body tissue with resulting biological damage. When you cook meat, you are effectively bombarding the meat with infrared radiation, the energy of this radiation being absorbed by the meat and changing it considerably. The radiations from radioactive substances can produce similar effects with massive exposures, but fortunately this sort of radiation burn is rare. The unit to measure this absorbed energy is the gray (Gy); 1 Gy is equivalent to one joule absorbed by one kilogram. (The gray supersedes the rad.)

The degree of damage that may be produced in biological systems by radiation is measured in terms of the rem. (At the time of writing there is a proposal to change to a new SI unit which will be called the sievert, abbreviated Sv. One sievert will be defined as an absorbed energy of 1 J/kg multiplied by a modifying factor dependent on the type of radiation. Since this unit has not yet been approved, we shall continue to use the rem as a unit of 'absorbed damage'.)

It is important to realise that we are all exposed to radiation all the time; we are being bombarded with a low level of radiation from outer space, there are radioactive minerals within the Earth's crust and even our bones contain small amounts of radioactive potassium and carbon. Whether these low levels are harmful or not is almost impossible to discover because there is no way to shield someone from them completely to provide a control experiment. One of the aims of radiological protection is to try to ensure that no one person receives a dose that is substantially greater than they would have received naturally. As a rough estimate, cosmic radiation, at sea level, will cause us to receive a dose of about 30 mrem per year (1 mrem = 1/1000 rem), whereas two hours' colour television viewing at a distance of two metres will cause you to receive a dose of up to about 1 mrem.

The ICRP (International Conference on Radiation Protection) is an international body that recommends maximum limits on the dose that any one person should receive. A classified worker, that is, someone who is subject to proper monitoring and medical checks, may not receive more than 5 rem in any one year. Pupils in schools or colleges (unless classified) should not be permitted to receive a dose in excess of 50 mrem per year; pupils under the age of 16 must not themselves perform any experiments involving the use of radioactive sources.

Sources of radiation may be classified into one of two types depending on how the radioactive material is presented. Sealed sources are those where the active material is securely contained in a metal capsule in such a way that it is only by the roughest handling that any will escape to the atmosphere. Open sources are sources where the active material is open to the environment and so there is a risk of it being ingested or inhaled.

There are three types of radiation that are given off by naturally occurring radioactive substances: called alpha, beta and gamma radiation; each has different properties and it is these properties that in many ways control the hazard associated with each.

Alpha particles, even those of high energy, only penetrate a few centimetres of air and are very easily stopped, even by a thin sheet of paper or cotton, and so present little danger to an experimental worker provided the active material is outside the body. Sealed alpha sources are therefore relatively harmless provided that they are handled with tongs to keep them a few centimetres away from the body. When not actually in use, they can be stored in any container which will then effectively contain the radiation. Open alpha sources present a different problem because, if swallowed or inhaled even in small quantities, considerable damage can result and special precautions need to be taken which may include full protective suiting.

Beta particles also are readily stopped by a metal sheet a millimetre or so thick. The comments made about alpha sources still apply.

Gamma radiation cannot be stopped but only reduced in intensity. The radiation is penetrating and can be detected, in the case of very active sources, at a distance of many metres. For safe working you should maintain as large a distance as is practicable from a gamma source or interpose lead brick shielding. A 10 mm thick lead brick will reduce the intensity of the radiation by approximately half and most laboratories use a combination of shielding and distance to reduce the dose that may be received. Considerable thought needs to be given to the arrangement of lead shielding walls since it must be remembered that gamma radiation will easily penetrate wooden bench tops or the brick wall separating you from the adjoining laboratory.

There are, then, three main factors that control the hazard to which someone working with radioactive substances is exposed. The first is the type of radiation that is being used; is it alpha, beta or gamma, an open source or sealed source? If it is an open source what sort of protective clothing should be worn? If it is a sealed source, is any shielding required and how far should you work from the source? The second factor is the dose rate that you are receiving; this must be kept as low as possible either by using shielding or moving away from the source and using remote handling equipment. The third factor is the length of time that you are exposed to the radiation. The total dose you receive is the number of rems per hour that you are exposed to multiplied by the number of hours you are exposed. This time must be kept as short as possible by working quickly. For some operations it may be possible to dramatically reduce the time it takes to perform an operation by practising with a dummy in advance.

Used properly, radioactive sources are perfectly safe and the dose received from them can be kept well within ICRP limits, but in the wrong hands they are potentially hazardous and should be treated with the same respect that is afforded to poisons. They should be kept, preferably locked, in a central store where a record must be kept of the sources that are held in stock and the date they were acquired. As these are taken out of store, they should be booked out so that a continuous record can be maintained of what stocks are held, what are in use and what have been disposed of in the authorised manner.

Any organisation wishing to make use of radioactive substances must obtain approval to do so from the relevant government department as laid down in the Radioactive Substances Act, 1960 and the subsequent Regulations for unsealed sources (1968) and sealed sources (1969). The authorisation that is given will lay down the type and strength of sources that may be used and the way in which they may be disposed of. Registration in this way also imposes on the employer the obligation to provide an adequate monitoring service to assess the dose received by each employee during any one year. There are various types of establishment that are exempt from registration under the 1960 Act; for example, schools and colleges using small sources only may apply to the Department of Education and Science for approval to keep and use them sources under the provisions of Administrative Memorandum 2/76 which imposes less onerous restraints.

6.3 X-RAY SOURCES

X-rays, although produced differently, have almost identical properties to gamma radiation. The only major difference is that X-rays can be turned off by switching off the generating set but gamma rays cannot. The actual dose received depends upon the radiographer, but each chest X-ray that an individual has will give him a dose of between 100 and 200 mrem.

X-rays are generated by accelerating electrons along a tube that has been pumped down to very low pressure, and allowing them to strike a metal target. They are brought to rest very quickly; some of their energy raises the temperature of the metal target, and some is given off in the form of X-radiation. The more energy that the electrons are given on their way along the evacuated tube, then the more penetrating are the X-rays produced. In this process the metal target gets extremely hot and in order to minimise damage it is usually rotated at high speed. On most machines, the whirring noise of this rotation is readily apparent during the exposure.

Oscilloscope and television tubes also accelerate electrons towards a target, the screen, and when they hit it, X-rays are also given off although at very low dose rate levels. Colour televisions give a higher dose rate than black and white sets because in colour sets the electrons are accelerated more vigorously.

Because of the danger of generating X-rays, care should be taken whenever a high voltage is applied to an evacuated tube; it should always be assumed that X-rays are being produced and appropriate safety measures taken. Special regulations of the Department of Education and Science apply to the use, in schools and colleges, of voltages above 5 kV.

A special word of warning is needed in connection with X-ray crystallography experiments. Beams of X-rays can emerge at totally unexpected angles and all such experimental set ups should be examined by a competent health physicist at regular intervals.

As was mentioned in connection with gamma radiation, it must be remembered that X-rays can easily penetrate walls, floors and ceilings and due consideration must be given to shielding not only an operator in the same room, but others working in adjacent rooms.

6.4 OPTICAL HAZARDS

The human eye is a sensitive and delicate organ, and whereas it can accommodate itself to cope with both darkness and bright sunlight, it must be protected from extremes of incident light. One would not, for example, look directly at the sun. In the laboratory there are two main types of extreme light condition: the first is the extremely intense beams of light that are produced by lasers (figure 6.2) the second is ultraviolet (u.v.) radiation.

Figure 6.2   Symbol to warn of laser hazard

Laser stands for Light Amplification by Stimulated Emission of Radiation. Lasers are capable of delivering a beam or a pulse of light that can be focused into a very narrow ray and thus concentrated on to a very small spot; the intensity of light on that spot is extremely high and generates considerable heat. Lasers have been used as cutting devices for cloth and metal, as drills for steel plate, as very fast acting dentist's drills and as a way of welding back into place the detached retina of a human eye. Such a concentration of energy must not be allowed to fall carelessly on the eye because quite obviously damage will result.

The source of a laser beam *must never* be looked at directly and one should not get into such a position that a beam might accidentally strike the eye after reflection off a wall or instrument casing. Ideally the path of a laser beam should be covered over and the target suitably shielded to prevent reflection and scattered light. When setting up an optical system using lasers suitable goggles should be worn, but it should be noted that different types of laser give different 'colours' of light, and the right type of goggles should be worn for the particular laser that is being used. A common type of continuous beam laser is one that is known as a helium/neon laser. This gives a red beam, and goggles that will not let this colour pass are needed.

There is an additional danger associated with lasers in that they require a very high-voltage source to operate. Experimental systems should be suitably covered to prevent accidental contact with live HT and commercial lasers should not be operated without the protective case except under the most carefully controlled conditions.

Work with lasers in schools and colleges is governed by the Code of Practice issued by the Department of Education and Science (Administrative Memorandum 7/70).

Sources of ultraviolet radiation are commonly used for causing fluorescence in some compounds, in some chromatography situations, for example. Ultraviolet light is also produced by carbon arcs and arc welding. Ultraviolet radiation is of shorter wavelength than ordinary light and more dangerous; overexposure can cause severe conjunctivitis. It has been suggested that for wavelengths that fall below 250 nm, conjunctivitis is almost certain.

When working with ultraviolet light proprietary goggles should be worn and these will give quite adequate protection though it should be noted that ordinary Pyrex glass will stop the more harmful rays and afford a reasonable measure of protection against all but the most powerful sources.

6.5 HIGH VOLTAGES

With any high-voltage system, there is the obvious danger of electrocution. As has been explained in chapter 2, accidental contact with a wire at or near to mains potential of 240 V may cause a current to flow from the point of contact to your feet or some other point that is at or near to earth potential (0 V). Quite often this current will be too small to detect because of the high value of the resistance in this path, particularly if you are wearing rubber-soled shoes. If instead of touching a wire at 240 V, however, you touch one at 5000 V, then the current will be about twenty times as large, which may well give rise to severe electric shock. All high-voltage connections should be made with the proper high-tension cable which is designed to cope with the maximum potential it may be asked to support. Connections should preferably be made with the insulated caps used with sparking plugs in cars, since these protect the whole connection from stray fingers. Finally a suitable warning sign should be placed as close to the HT connections as possible (figure 6.3). When dealing with very high voltages, it should be remembered that

Figure 6.3 Symbol to warn of high-voltage hazard

sparks can jump across air gaps and there is a possibility of a spark jumping to find earth through the human body if you come too close to a conductor at high potential; it needs approximately thirty thousand volts (30 kV) to jump a gap of 1 cm.

As has been mentioned previously in this chapter, voltages in excess of about 5 kV applied to evacuated tubes can generate X-rays and care should be taken to see that this doesn't happen unexpectedly.

In applications where very little current is to be drawn from the HT supply, for example, in electron beam deflection systems, it is advisable to connect a high-value resistor in series with the supply terminals. A 5 megohm ($5\,M\Omega$) resistor might be suitable; this will limit the maximum current that can be drawn from the supply thus minimising the electrocution risk. In many of the high-tension supplies that are marketed for use in schools, this is done internally for safety reasons.

6.6 COMPRESSED AIR AND VACUUM SYSTEMS

Many workshops and laboratories have a piped-in supply of compressed air run from a compressor sited centrally within the building. For convenience such supplies are usually run at relatively high pressure and, if used improperly, can be extremely dangerous, particularly as they seem at first sight to be so harmless. Appalling injuries can be caused by directing a com-

pressed air line at someone's face and this simply must never be done. It is not good practice to direct an air line at the body at all because there is a risk, particularly if someone has a cut or abrasion, of forcing air into that person's blood stream. For the same reason, you should never use a compressed air line for drying your hands. Because of the pressure at which most systems are run, they should not be used for drying samples or cleaning dirt and dust from apparatus; they are more likely to blow objects across the room and cause damage rather than act as a quick and simple cleaning mechanism.

The main danger to an operator from a vacuum system is the possibility of implosion of a glass vessel when the pressure inside is suddenly reduced. Large glass containers such as bell jars should be evacuated slowly and progressively, first with the rotary pump only and then by a diffusion pump. They should always be fitted with a wire mesh guard that will stop pieces of glass flying around the laboratory in the event of an accident. If wire guards are not available, then clear adhesive tape wound round at regular intervals has been found to be an acceptable substitute.

6.7 CHEMICAL HAZARDS

So often in physics laboratories chemicals are not treated with sufficient caution, even by workers who would themselves take a quite different attitude in an adjacent chemistry laboratory. It must be emphasised that whenever someone is handling chemicals they should wear the appropriate safety equipment. There are still some laboratories which use chromic acid techniques for cleaning glassware; when handling chromic acid it is desirable to wear a laboratory coat, rubber gloves and some sort of full face protection.

Mercury vapour is poisonous and mercury should not be left around open to the atmosphere and spillages should be treated immediately (even small quantities can be recovered from the floor using a teat pipette almost horizontally) although working within a spillage tray should eliminate this sort of accident.

6.8 ASSIGNMENTS

6.1 In your own area of work, what equipment is left running overnight or during holiday periods? What automatic cut-outs are there in case of emergency?

6.2 If your work place uses radioactive sources, what systems are in operation to ensure that a small quantity of active material does not get left accidentally in someone's pocket?

6.3 What is the activity (in becquerel) of the largest radioactive source your organisation deals with? What is the dose rate 1 m away from this source?

6.4 If your establishment uses radioactive sources, is it exempt from, or under the control of, the Radioactive Substances Act, 1960? If it is under this control, where is the certificate of registration?

6.5 What is the commonly accepted warning sign for laser light?

6.6 Which warning notices are fitted to your own organisation's compressed air system?

7 Safety in Biology Laboratories

7.1 GENERAL RULES

Biology laboratories may contain a wider range of hazards than any other laboratory. A well-equipped laboratory will frequently contain much electrical instrumentation and will certainly possess a wide variety of chemicals. It is therefore important to read chapters 5 and 6 on hazards in chemistry and physics laboratories first; the contents of this chapter explain additional hazards.

Perhaps the most important aspect of safety is the technician's attitude towards it. Accidents often happen when a situation is considered safe; obvious dangers tend to induce more care and attention. In this respect, one of the greatest dangers in a biology laboratory is work with micro-organisms — because they are invisible the dangers are not obvious, so sufficient care may not be taken. A technician must be on guard not only with obvious hazards, such as dissection equipment, but with all laboratory activities at all times (figure 7.1). An understanding of potential problems will help to prevent their occurrence.

BIOHAZARD

Figure 7.1 Symbol to warn of biological hazard

As in other laboratories, the reporting of accidents is very important. In biology laboratories, sickness must also be recorded, because this could be an indication of a disease having been contracted in the work place. All skin injuries should also be notified, particularly if working with micro-organisms. When reporting illness to a doctor, it will be helpful to him if you mention the materials with which you have been working, and also keep your safety officer informed.

The basic rules are

(1) take care at all times — hazards are not always obvious

(2) avoid rapid movements and activities

(3) beware of sharp instruments like hypodermic needles, scalpel blades, etc., and abrasive materials such as ice; even small skin punctures can inoculate you with a pathogen (disease-causing organism)

(4) wear suitable protective clothing at all times; this should never be less than a laboratory coat, and special situations may demand more extensive protection

(5) observe strict personal hygiene, hand-washing, etc.

(6) In the laboratory never let anything come into contact with the mouth because this is another source of entry for pathogens; this prohibition includes eating, drinking, smoking, pencil-sucking and nail-biting

(7) keep the safety officer informed of all accidents and illnesses.

These are not the only rules, as you will see later in the chapter, and some are more essential in one laboratory than another, but they are all always important.

7.2 ADDITIONAL CHEMICAL HAZARDS

Because a wide variety of chemicals is now found in biology laboratories, all the points mentioned in chapter 5 are also relevant here, but biological situations also possess additional hazards, either because they use chemicals not usually found elsewhere, or because they are used in larger quantities or under different circumstances. Dangerous chemicals may be classified as follows

(1) toxic

(2) flammable

(3) explosive

(4) corrosive.

Toxic chemicals include carcinogens — that is, substances which tend to produce cancer — and teratogens, which tend to produce mutations in unborn children.

7.2.1 Carcinogens

Great care must be taken with suspected carcinogens; because there is sometimes a long time interval between exposure and effects, the dangers are far from obvious, which makes it all too easy to take insufficient care. Evidence is coming to light to show that, in some cases, quite small exposure to such materials may increase the chances of cancer, sometimes many years after contact. This makes the identification of these chemicals slow and difficult, which means that we may now be handling chemicals with unknown carcinogenic activity.

Carcinogenic substances which may be found in a biology laboratory include, for example, benzene, sometimes used as an organic solvent, and ninhydrin, used in the detection of amino acids and proteins. Ninhydrin should never be used from an aerosol spray. Formalin reacts with any hydrogen chloride in the air, or with the hydrochloric acid in the stomach, to form a compound which is possibly carcinogenic.

7.2.2 Teratogens

These include anti-tumour drugs, cortisone and thyroxine. Certain herbicides may also be teratogenic.

7.2.3 Other Toxins

Some substances normally present in the human body are toxic in excess. This applies to many vitamins, enzymes and hormones.

There are several wild and garden plants that are poisonous to varying degrees when ingested. They may contain toxic alkaloids and glycosides. Some examples are monkshood (*Aconitum* spp.), foxglove (*Digitalis purpurea*), deadly nightshade (*Atropa belladonna*), woody nightshade (*Solanum dulcamara*), death-cap fungus (*Amanita phalloides*) and laburnum (*Laburnum anagyroides*). Other plants can cause dermatitis, for example, cowslips (*Primula veris*), daffodils (*Narcissus* spp.) and, particularly, the greenhouse plant *Primula obeonica*. A number of seeds may be quite poisonous, and the ingestion of castor-oil seeds can be lethal.

Pesticides are chemicals used to kill undesirable organisms and include insecticides, herbicides, fungicides and rodenticides. Most of these are also toxic to man. Many can enter the body through the skin so that all skin contact with them should be avoided. Examples of poisonous insecticides are DDT—a chlorinated hydrocarbon, which is persistent, that is, it is not broken down in the environment—and malathion, which, although not persistent in the environment, will, like DDT, accumulate in the tissues over a period of many years if ingested. Insecticides and fungicides are potentially most dangerous when used for greenhouse fumigation, when care should be taken to follow exactly the instructions given with the product. Among herbicides, paraquat has caused a number of deaths of children.

7.2.4 Disinfectants

Disinfectants are much used in biology laboratories and animal houses, or in any laboratory where there may be pathogenic organisms. Those based on phenol, for example, Lysol and Sudol, are toxic and are skin irritants; they must be used with care, especially when undiluted. Other common disinfectants, such as the hypochlorite solutions, are less toxic.

Flammable chemicals found in biology laboratories include those mentioned in chapter 5. Some anaesthetics are flammable, for example, ether, which is both flammable and explosive, and trichlorethylene (Trilene) which is flammable and forms toxic substances with soda-lime. Probably the least hazardous way of killing small vertebrates is by using carbon dioxide from a cylinder introduced into an approved killing chamber.

Explosive chemicals include ether, as has already been mentioned, and picric acid, used in many histological preparations, is explosive when dry. For this reason picric acid should never be stored in bottles with ground glass stoppers, because it may crystallise around the stopper and explode as the stopper is loosened.

7.3 HAZARDS INVOLVING EQUIPMENT AND APPARATUS

Many pieces of equipment used in biology laboratories have a mains electricity power supply. In addition some equipment, such as electron microscopes and electrophoresis apparatus, has a high-tension supply.

The passage of an electric current through the body may result in shock, burns and even death. The size of the current that will cause injury varies from person to person and also depends on whether the supply is a.c or d.c. The parts of the body through which the current flows are also relevant. (See table 7.1.)

Table 7.1

	50 Hz a.c.	d.c.
Threshold of feeling	1 mA	5 mA
Muscular paralysis	15 mA	75 mA
Ventricular fibrillation (heart stops)	70 mA	300 mA

The voltages commonly used in electrical apparatus are potentially lethal, particularly so if contact is made through a low-resistance connection. Skin moistened with saline or any physiological solution will have such a low resistance.

Centrifuges are all dangerous if operated carelessly. To prevent the formation of aerosols, the tubes should never be more than two-thirds full. If pathogenic material has been used the head and bowl should be disinfected, but not with a hypochlorite disinfectant because this is corrosive, and corrosion is a major source of rotor failure.

Autoclaves must be submitted to a regular hydraulic test by a qualified engineer, and should be tested tegularly at specified time intervals. The pressure release valve should be set by the engineer, and must not be adjusted to a higher setting than that specified by the manufacturer.

Care must always be taken when using an autoclave. Fill it to the correct level with water, avoid overloading, make sure that the openings to the pressure gauge and safety valves are not blocked, and do not open it until the pressure has dropped to atmospheric pressure. The contents will be extremely hot, so use heat-resistant gloves when unloading. Screw-cap bottles must never be autoclaved with their tops screwed on, because the glass may not be able to withstand the changes in pressure.

Refrigerators are potentially dangerous if used to cool volatile, flammable organic solvents, as might be required, for example, for chromatography of labile substances. As mentioned in chapter 5, only refrigerators fitted with a flash-proof thermostat can be used for this purpose, or fire and explosion may possibly result.

Refrigerators can often become a repository for old and forgotten vessels. To avoid this, all containers should be securely stoppered and clearly labelled, showing the nature of the contents, the date and the name of the person responsible for placing that vessel there. A container of any known pathogenic or toxic material should stand in its own tray, to avoid contamination of other vessels and to aid disposal and collection in the event of its being cracked or broken.

No food stuffs for human or animal consumption should ever be stored in a laboratory refrigerator.

7.4 SAFETY IN MICROBIOLOGY AND PATHOLOGY LABORATORIES

The basic safety rule here is to assume that any specimen or culture which has not been sterilised contains pathogens. This section is applicable to microbiological work, work with pathology and histology specimens which have not been fixed and work with blood in transfusion centres and haematology laboratories.

It is worth noting that the Department of Education and Science, in *The Use of Micro-organisms in Schools* (1977), has named eight organisms (*Chromobacterium violaceum, Clostridium perfrigens, Pseudomonas aeruginosa, Pseudomonas solanaceaerum, Pseudomonas tabacci, Serratia marcescens, Staphylococcus aureus, Xanthomonas phaseoli*) which are now considered to be unsuitable for use with pupils below the age of 16. These have all been recommended previously in certain Schools Council and Nuffield science curriculum projects. Bacteria should not be painted on the skin or sprayed up the nose as had been suggested. Due to the dangers of mutation of the bacteria into more dangerous forms, of allergic reactions and of difficulties of distinguishing some species from similar but dangerous varieties the Department of Education and Science emphasise the need to regard all cultures as potentially hazardous.

The micro-organisms which live in or on the human body are of two types: commensal organisms which do no harm and pathogenic organisms which cause disease. The distinction between them is not clear cut, and a number of commensal organisms can be pathogenic if introduced and allowed to multiply in a habitat other than their normal one. Bacteria normally living harmlessly in the intestines, for example, *E. coli*, can cause septicaemia if present in the blood stream. Personal hygiene and the prevention of cuts and abrasions to the skin are obviously very important here.

Safe work in the microbiology laboratory may be considered at two levels. The first is the containment of micro-organisms, keeping them away from human contact. The second is the protection of the technician who handles micro-organisms.

7.4.1 Containment of Micro-organisms

The standard techniques for handling micro-organisms are designed to minimise contact between the operator and the organisms; they must be applied rigorously at all times so that they become automatic. There is not sufficient space here to list them in detail, but reference should be made to any textbook on practical microbiology.

If known dangerous pathogens are to be handled, then special containment techniques must be used. When handling organisms in the open laboratory, however, it is almost impossible to contain them completely. One of the commonest dangers is the formation of aerosols, that is, fine solid or liquid suspensions in air. The particles or droplets are so small that they are invisible, and may be carried in the air for a long time. If breathed in, they may reach the lungs. Being invisible, it is difficult to appreciate how easily aerosols are formed, and how persistent they are. Any operation that breaks a liquid surface will cause some aerosol formation; the following list suggests that it is almost impossible to prevent aerosol formation in the open laboratory

(1) opening screw-caps, snap-on closures and plug stoppers

(2) opening ampoules of lyophilised material

(3) rinsing pipettes and other glassware, or expelling the contents of pipettes and hypodermic needles

(4) using mechanical homogenising or ultrasonic apparatus

(5) using mechanical spraying apparatus

(6) sputtering when sterilising a loop

(7) grinding dry materials

(8) expiration by animals, especially sneezing

(9) incorrect use of a centrifuge

(10) accidental breakage of containers of culture.

When using any known dangerous pathogenic material and any pathogen of the respiratory passages, special precautions must be taken to avoid contact between aerosols and operator. The most important aspect of this is the use of an exhaust protective cabinet. The principle of this apparatus is that air is constantly being drawn into it from the laboratory. Any aerosols produced when working with cultures inside it are drawn with this air flow through fine filters, capable of removing all aerosols, and then exhausted out of the building. If the air flow is great enough, there should be no danger of introducing an aerosol into the air of the laboratory. When not in use, the inside surfaces are sterilised by ultraviolet light, but the apparatus should also be disinfected with chemical disinfectants.

If a Petri dish or other container holding pathogenic bacteria, fungi, tissue cultures, virus cultures or other pathological specimens is accidentally overturned or broken, the following procedure should be adopted.

(1) Stop working and evacuate the area.

(2) Open the nearest window and allow at least 15 minutes for aerosols to disperse.

(3) Put on appropriate protective clothing, and flood the spilt material with disinfectant, leaving it for up to an hour.

(4) Pick out any broken glass with forceps, and place in a disinfectant bath.

(5) When disinfection is complete, wash the area with water.

(6) If skin or clothes have been contaminated, disinfect skin with chlorohexidine and autoclave clothing.

(7) Report the accident and record the incident in the accident register or whatever system is used.

The opening of ampoules of lyophilised cultures is hazardous unless performed correctly. The contents are liable to be dispersed into the atmosphere, especially if the container has been sealed under vacuum, or is exposed to draughts. To avoid this, the following procedure is recommended.

(1) File a small mark at about the middle of the cotton-wool plug and apply a red-hot glass rod to crack the glass of the ampoule.

(2) Allow sufficient time for air to filter into the ampoule.

(3) Put on rubber gloves, wrap a disinfectant-soaked swab around the neck of the ampoule and carefully break off the pointed end.

(4) Remove the plug with forceps, place it and the broken off end of the ampoule into disinfectant.

(5) Flame the end of the ampoule, and insert a sterile plug of cotton wool.

The ampoule is now ready for the sterile broth to be added to re-suspend the culture.

7.4.2 Protection

Assuming that containment techniques are never completely effective, the technician working with micro-organisms must take certain precautions to protect himself from infection. It is important to be aware of the variety of routes by which micro-organisms can enter the body

(1) through the alimentary canal, following ingestion, for example, gastroenteritis

(2) through the mucus membrane of the nasopharynx, for example, the common cold, influenza

(3) through the mucus membrane of the lungs, for example, tuberculosis

(4) through the mucus membrane of the urino-genital system, for example, gonorrhoea

(5) through the skin, following cuts and abrasions, for example, tetanus and rabies.

Certain basic precautions can be taken to lower the risk of entry of micro-organisms into the body. The standard protection of a laboratory coat is useful, but not unless it is worn properly. It should be completely buttoned up or, preferably, the surgeon style of coat should be worn. The coat must be sterilised regularly and not allowed to leave the laboratory.

Long hair is very easily contaminated, and is very difficult to sterilise; it should therefore be tied back safely.

Infection through the alimentary canal follows ingestion. This usually occurs accidentally, when the hands or some object becomes contaminated, and these are then allowed to come into contact with the mouth. The laboratory rules of prohibiting eating, drinking and smoking are designed to prevent this. It is essential that you thoroughly scrub your hands, after taking off your laboratory coat, and before leaving the laboratory area.

Hands may become contaminated by touching almost any solid surface in the laboratory. For this reason, the minimum precaution is to disinfect the bench top with a detergent-based agent, such as Sudol, at the end of every day. (A disinfectant is a chemical which reduces the number of infectious organisms to a safe level; for complete elimination of micro-organisms, sterilisation must be carried out.) Refrigerators and other apparatus should be regularly cleaned in the same way. A disposable, plastic-backed, absorbent bench covering, such as Benchkote, reduces the need for bench disinfection.

Spillages are less likely if the bench working area is uncluttered. It is inadvisable to write notes in the immediate area where you are working, because paper and writing instruments are not designed to be sterilised. Self-adhesive labels must always be used in preference to those that need to be licked and pipetting should never be done by mouth. Old cultures and possibly contaminated apparatus should never be left around but must be discarded into

disinfectant such as Sudol or Chloros which is prepared freshly every day. Contaminated liquids are safely disposed of by mixing them with an equal volume of disinfectant.

Infection by penetration of the skin should be prevented by taking care with hypodermic syringes, sharp instruments and broken glass which should never be picked up by hand. Chipped or scratched glassware should be discarded, because this is liable to break following rough handling. Any known scratches or abrasions on the skin should be disinfected with a suitable skin disinfectant such as Hibitane, which contains chlorohexidine, and covered with a dressing.

Technicians, especially those working in microbiology or pathology laboratories or animal houses should be fully protected by immunisation. Advice on this will be given by the safety officer.

7.4.3 Handling Human Blood

In all educational laboratories, the taking of blood samples from students must be avoided. When samples are required, the nearest blood transfusion centre should be contacted.

When using human blood, the same safety procedures should be adopted as for a microbiology laboratory, even if the blood is supplied by a transfusion centre. The reason for this is the risk of contracting hepatitis, a condition which is difficult to treat and from which recovery is very slow.

There are two diseases which cause the symptoms of hepatitis, one caused by virus A (infectious hepatitis) and one by virus B (serum hepatitis, post-transfusion hepatitis, inoculation jaundice). Virus B is the cause of danger when using human blood in the laboratory. Although blood transfusion centres screen blood donors to detect carriers of the virus, who may show no symptom of the disease, they cannot test every blood pack they receive, so there is a small risk associated with any blood sample, no matter where it is obtained. The disease can be contracted when blood containing the virus comes into contact with even slightly damaged skin or mucosa.

7.5 SAFETY WITH ANIMALS IN LABORATORIES AND ANIMAL HOUSES

In school biology laboratories, it is most important to stress that wild animals, dead or alive, must never be brought into the laboratory. They are invariably verminous, as children's pets may be. Animals raised in animal houses are less likely to carry vermin or disease, but are not free from them.

It is important to remember that the housing and use of animals for experimental purposes is subject to both statute law and regulations issued under the authority of statutes. Some of these concern safety aspects. Sources of information on these matters are listed at the end of the book.

Some of the possible dangers involved with the use of animals in the laboratory include

(1) physical injury, scratches, bites, etc.
(2) infection by bacteria, viruses or fungi, either from the animals themselves (zoonoses) or from the bedding or cages in which they are housed
(3) attack by parasites; ectoparasites, for example, fleas and lice, can be transmitted directly by contact; endoparasites, for example, flukes and tapeworm, can be transmitted by ingestion of eggs or other stages, or penetration of the skin by specialised stages of the parasite's life-cycle
(4) poisoning by venomous animals
(5) allergic reactions.

The avoidance of scratches and bites must be achieved by using the correct methods of handling each type of animal. Where animals are regularly used, a qualified animal technician should be available and responsible for the animals.

The prevention of infection to the handler when in contact with animals poses similar problems to the prevention of infection from isolated micro-organisms.

A scratch or bite is a very efficient method of inoculation, and this is the principal method of infection by rabies, which may at any time become established in the United Kingdom. Any bite, scratch or cut which has been in contact with animals must be treated immediately, and an entry made in the accident book or register. The wound must be thoroughly cleaned with soap and water, and bleeding encouraged. The wound should then be treated with a skin disinfectant and a protective covering applied. All injuries should be seen on the same day by a doctor, who should ensure that the person is protected against tetanus.

The prevention of infection by zoonoses starts with typical microbiology laboratory precautions such as personal hygiene, avoiding contact with the mouth, careful working, the wearing of protective clothing, etc. Unless working in a specific pathogen-free (SPF) unit, the environment is far from sterile, but good animal house hygiene can help to reduce the risks.

All cages should be thoroughly cleaned and either disinfected or sterilised on a regular rota basis. The cages must be designed for easy cleaning and sterilising, and be secure. Used bedding should be put into waterproof containers for disposal, and autoclaved or incinerated.

An infected animal should be immediately isolated or destroyed, and the carcass incinerated. The bedding and cage must be sterilised.

Many zoonoses are transmissible from apparently healthy animals. As in a microbiology laboratory, there is considerable danger from invisible aerosols containing pathogens. Fumigation may be the only way of reducing the risk.

Perhaps the most dangerous group of zoonoses is that carried by primates; for this reason, monkeys should never be used in school laboratories. The two main hazards to handlers are B virus and tuberculosis; monkeys can also transmit rabies. Dysentery is also a common cause of mortality among imported monkeys, and the faeces of even apparently healthy monkeys have been shown at times to contain a variety of potential pathogens.

Laboratory rats may harbour some species of *Salmonella*, which may become established in the colony without noticeably affecting mortality. Humans may become infected by eating food contaminated by the organisms, resulting in a type of gastroenteritis with symptoms of fever and diarrhoea.

Members of the parrot family may transmit psittacosis to man. Many species of mites, fleas, lice and ticks will attack humans in order to suck blood, and they are often extremely efficient at transferring themselves from an animal.

There is a wide variety of zoonose endoparasites in the following groups: protozoa, nematodes, trematodes (flukes) and cestodes (tapeworms), which are often easily transferred to humans, and from which no animal may be assumed to be free.

Freshly killed animals kept in a freezer are almost as hazardous as when alive. The micro-organisms in them may be just as viable and, following a prolonged thaw, may increase greatly in numbers.

Other methods of preservation also present problems. Before preservation the carcasses should be fixed, a process which halts bacterial putrefaction and denatures enzymes. It is worth remembering that fixatives have similar effects on the living tissue of the technician. It is also not fully realised that a fixative may not totally sterilise the tissues treated.

Although it is widely realised that many reptiles are venomous, it is not so well known that there are many other venomous animals, particularly marine invertebrates and fish.

Allergic reactions such as anaphylactic shock can be severe enough to cause death, and so care must be taken with the causative agent. A bee or wasp sting, for example, can kill a sensitised person. It is not uncommon to find that some individuals become sensitised to asthma when working in close proximity to large numbers of locusts.

7.6 EXPERIMENTS ON HUMANS

In educational establishments, it may sometimes appear to be necessary to perform simple experiments on pupils or staff. The people concerned must be over 16, have given their full consent, and be in full knowledge of the possible hazards associated with the experiment.

Phenylthiourea tasting should not be performed, nor any chemical swallowed.

Great care should be taken with spirometry experiments involving unusual lung ventilation, or following unusual physical stress. Such experiments can be dangerous to epileptics, and people suffering from bronchial disorders, which may not be known.

Stroboscopic lights at certain frequencies can induce epileptic fits in persons not normally considered epileptic. Similarly, EEG (electroencephalograph) biofeedback experiments should be avoided because of possible unwanted mental effects.

7.7 ASSIGNMENTS

7.1 List the biological materials that present the major hazards in your laboratory either at college or at work. What precautions need to be taken to minimise these risks?

7.2 What regulations are in force, either at work or college, to ensure that hazardous materials are left in the laboratory and not taken out with you when you leave?

7.3 What protection by means of inoculation, etc., is available for work in animal houses?

7.4 Has the local hospital or site doctor been informed of the types of biological hazard to which staff are potentially exposed?

8 Scientific Measurement and Reporting

8.1 ACCURACY

Whenever a measurement is made and the result noted, it is important that a realistic estimate of the accuracy of that measurement is also recorded. In general terms, if a measurement to a high degree of accuracy is required, it takes a long time to make and often requires expensive and delicate equipment. In many situations a very high degree of accuracy is unnecessary and a measurement may be taken more quickly, easily and cheaply but it is essential to indicate what degree of accuracy was aimed for. If you do not, there is a real danger that at some time in the future, someone may assume that your rough approximation was an accurate result, and, basing some calculation on this misconception, claim some result that is quite unjustified.

The accuracy of a result is usually best expressed by specifying a band between which you are sure the real answer lies. Suppose, for example, you are asked to measure the length of the piece of wood in figure 8.1 using a ruler. As you can see, the length of the wood is not an exact number of the divisions on the ruler. It is possible to try to guess how far between the divisions it actually lies but at best your guess can only be a rough estimate. What you are sure of, however, is that the length of the wood is somewhere between 21 mm and 22 mm. For convenience, this is usually written as 21.5 ± 0.5 mm; if you subtract 0.5 from 21.5, you get the lower limit that you are sure of, whereas if you add 0.5 to 21.5 you get the upper limit that you are sure of. 21.5 ± 0.5 is therefore a way of specifying a range of values between which you are sure the answer lies even though you cannot be certain where in this range it lies exactly.

It is important to specify the accuracy of any measurement that you make whether it be of length, mass, temperature, pH value or volume injected into an animal using a hypodermic syringe. In the remainder of this chapter, some examples will be given of common measuring devices and the degree of accuracy of which they are capable.

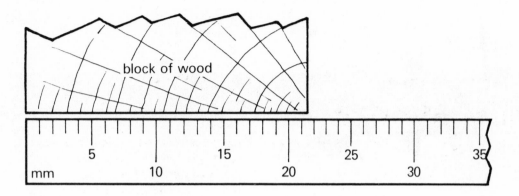

Figure 8.1

8.2 VERNIER CALLIPERS

There are many different makes and types of callipers on the market, some capable of being read to a greater accuracy than others. We shall describe one of the cheaper types for the sake of simplicity (figure 8.2).

The object to be measured is either placed within the 'outside' jaws, which are then gently closed until they just touch the sides of the object, or around the 'inside' jaws which are opened until they just touch the inner surfaces of the object to be measured. Many callipers also offer the facility for measuring the depth of a hole by using the centre slide (figure 8.2c).

The scale that can be seen on the sliding jaw in figure 8.3 is called a vernier scale, and this one allows the opening of the jaws to be measured accurately to ± 0.1 mm. The vernier scale is exactly 9 mm long, but is divided into 10 equal divisions. Each one of these divisions is, therefore, 0.9 mm long (figure 8.4). Imagine sliding the vernier scale along the main scale until the first vernier line corresponds to the first line on the main scale (figure 8.5). It can be seen that the distance x, which is the amount by which the jaws have been opened, is equal to 0.1 mm. If the vernier scale had been moved so that it was the second line on the vernier which lined up with the second line on the main scale, then the jaws would have been opened 0.2 mm. In figure 8.6, you can see that the fifth

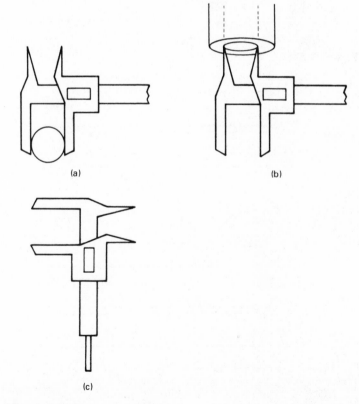

Figure 8.2   Vernier **callipers**

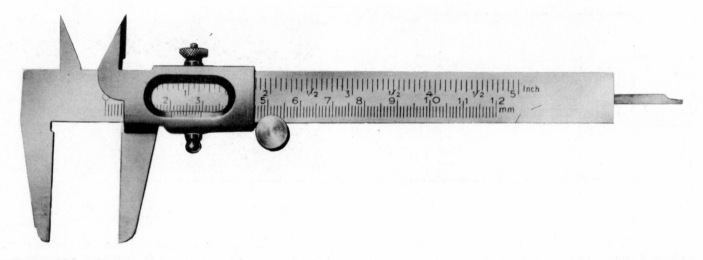

Figure 8.3

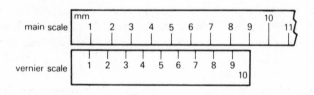

Figure 8.4

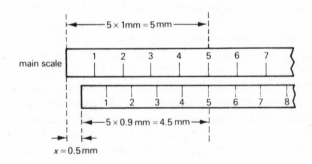

Figure 8.6

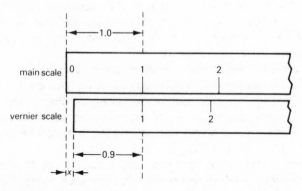

Figure 8.5

line on the vernier scale corresponds to the fifth line on the main scale making the gap x equal to 0.5 mm.

As a general rule with this type of callipers, the number of tenths of millimetres that must be included in your measurement can be found by looking at the vernier scale and noting which vernier line exactly coincides with a line on the main scale; if it is the fifth, you add 0.5 mm, if it is the sixth, you add 0.6 mm, if it is the eighth, you add 0.8 mm, and so on.

Figure 8.7 shows a pair of vernier callipers being used to

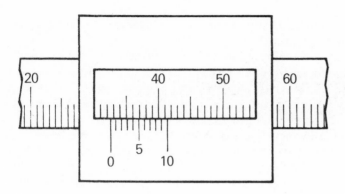

Figure 8.7

Figure 8.8 Micrometer screw gauge with digital readout

measure the external diameter of a cylinder. The callipers are read by first counting the number of whole millimetres to the zero of the vernier scale (in this case 32) and then looking at the lines on the vernier to see which one lines up with a line on the main scale (in this case the fourth line) so 0.4 mm must be added to 32 mm to give the diameter of the cylinder as 32.4 mm.

In practice it is most unlikely that one of the vernier lines will exactly line up with one on the main scale and it may be that you are unable to decide which vernier line to choose. For example, the number of whole millimetres might be 32, but you are unable to decide between the third, fourth and fifth vernier lines. In this case, you are sure that the true measurement is somewhere in the range 32.3 mm to 32.5 mm, and so this result can be most easily expressed as 32.4 ± 0.1 mm which exactly covers the band in question.

Verniers are not only fitted to callipers but to a wide variety of measuring devices, including travelling microscopes, spectrometers, potentiometers and some simple balances.

8.3 MICROMETER SCREW GAUGES

Micrometers are widely used where measurements of thickness or other external dimensions are required to an accuracy greater than can be obtained with vernier callipers (see figure 8.8). Even with a simple gauge, measurements can be taken which are accurate to 0.01 mm. The spindle screws in and out of the frame in such a way that for every complete turn, the end of the spindle moves exactly 0.5 mm. Fifty marks are equally spaced around the outside of the spindle. Turning the spindle through one complete turn (0.5 mm) means moving 50 of these marks past the reference line on the sleeve; moving one mark past this reference line corresponds, therefore, to a movement of the end of the spindle of 0.5/50 mm = 0.01 mm.

To read the micrometer shown in figure 8.8 it is necessary first to count up the number of millimetres and half millimetres shown on the sleeve (in this case it is 3.5 mm); the twenty-first micrometer line is against the reference mark on the sleeve, and so 0.21 mm must be added to the reading. In this case, the micrometer is measuring a distance of 3.5 mm + 0.21 mm which is equal to 3.71 mm.

Occasionally, usually as a result of careless handling, it will be found that with the micrometer just closed, the zero mark on the sleeve does not line up so that the micrometer reading implies that there is an opening of 0.01 or 0.02 mm when in fact the opening is zero. If you need to take a reading in a hurry, note whether the micrometer is over-reading or under-reading, and then add or subtract this zero error from your final measurement. With most micrometers it is a short task to reset the instrument so that there is no zero error. Instructions on how to do this are invariably supplied by the manufacturer when the instrument is bought.

The presence of a zero error is a sure sign of careless handling which most commonly takes the form of over-tightening the screw; most micrometers provide a ratchet which enables you to tighten the spindle on to the object being measured, but which is designed to slip as soon as the spindle comes into contact with an object thus preventing over-tightening.

In practice it is uncommon to find one of the marks on the spindle exactly lining up, and you may well find yourself unable to be sure whether a reading should be 3.71 or 3.72 mm; as before, this can easily be expressed as 3.715 ± 0.005 mm thus specifying the range within which you know the true reading lies.

More sophisticated micrometers provide a vernier scale to enable you to estimate a third decimal place, but it is our experience that only skilled craftsmen who have grown used to a particular instrument over a period of time are able to gain much benefit from this.

At the time of writing a new generation of electronic digital micrometers has just been launched at about three times the cost of conventional designs and, as yet, unproved in reliability over a period of years.

8.4 BALANCES

Recent years have seen a revolution in laboratory balances; the cumbersome beam balances that were in common use even in the 1960s have now almost all been replaced with electromechanical or electronic systems. For general-purpose laboratory use top pan balances are almost universally used. With these, the object whose mass is required is placed on a pan at the top of the balance which is depressed by an amount proportional to the unknown mass. This depression is sensed by one of a variety of mechanisms; one of these notes the change in resistance of a strain gauge attached to a bending beam and the unknown mass is displayed either electronically on a digital readout, or mechanically on a moving scale. A typical example of this type of balance would be capable of weighing up to a maximum of 1 or 1.5 kg to an accuracy of ± 0.01 g. As a general rule, balances which are capable of

weighing larger over-all masses are not as sensitive as those which have an upper limit of, say, 500 g.

As with all balances of this type, the major precautions that must be taken before use are to see that the balance is level (often a spirit level is built into the casing and levelling can be carried out by means of adjustable feet) and on zero before use. Most balances of this type will have a knob with which the zero can be set.

Ideally these balances should be placed on a firm bench top or other immovable support; once levelled, they should not be moved unless absolutely necessary.

More accurate balances are sometimes required, particularly in analytical chemistry. These often have a suspended weighing pan in an enclosed space to eliminate draughts; the object to be weighed is placed on the pan and the lid or door then closed. This type of balance often uses a beam with small rings of precisely known mass being lifted off the end of the beam which carries the load. When the mass of the rings which have been lifted is equal to the mass on the pan, equilibrium will be restored; this may be indicated by a dial or meter on the front of the balance. The settings of the control knobs which lift the rings will enable the unknown mass to be determined. This arrangement has the advantage of keeping a constant load on the pivot of the beam.

Balances of this type can frequently weigh up to a mass of 300 g with an accuracy of ± 0.001 g. When determining mass to such an accuracy as this, vibrations transmitted through the floor can cause difficulties and it is usually preferable to mount this type of balance on a firm concrete support. Once installed this type is never moved except for maintenance.

All balances should be checked regularly against standard masses to ensure that they are still accurately calibrated (see chapter 9). Except in the case of the largest research establishments, it is usual to leave the maintenance and servicing of this type of balance to an external firm of specialists or to the suppliers.

8.5 VOLUMETRIC GLASSWARE

Quantitative analysis in chemistry frequently requires the accurate measurement of volumes of solutions or distilled water to make up standard solutions and a wide range of graduated glassware is available for this. Volumetric flasks have a narrow neck around which is a scratch mark. If the flask is filled to this mark, it will contain the quantity of liquid that is specified on the flask, provided that it is at the specified temperature, usually 20 °C. The flask should be filled so that the bottom of the meniscus just lines up with the scratch mark. Your eye must be on the same horizontal level as the scratch mark to avoid parallax errors. Provided that the solution you are making up is within about 5 °C of the stated temperature, the volume will be accurate within the limits stated by the manufacturer.

Pipettes are a cheap and convenient way of dispensing a known volume of liquid. Standard-style pipettes are available in many sizes ranging from about 5 cm^3 up to 500 cm^3 although other sizes are available for more specialised applications. Again they are calibrated so that when full with the bottom of the meniscus just level with the scratch mark they then contain whatever volume is engraved on their side, provided that the liquid is within a few degrees of the temperature at which the calibration was made. This is generally 20 °C, but should always be marked on the glass. Different quality pipettes are available (at different prices) with differing standards of accuracy, grade 'A' being more accurate than grade 'B'.

It is unwise to suck anything up a pipette by mouth, except perhaps distilled water, and there is a variety of inexpensive devices on the market which allow an operator to use a pipette quite safely. One such device is a rubber bulb with valves to allow liquid to be sucked up beyond the scratch mark, expelled drop by drop until the meniscus lines up with that mark and then expelled rapidly into whatever container is appropriate.

Remember that, when filling, the tip of the pipette should be kept well under the surface of the liquid, and that when expelling the measured contents, a small drop will remain behind. This has been allowed for in the design and calibration and the pipette should not be shaken or blown through in an attempt to get it out.

The accuracy with which a stated volume of liquid can be dispensed using a pipette depends on two main factors. The first is the quality of the calibration and the second is the efficiency of the operator in filling it exactly to the scratch mark. This is a skill which comes with practice but in the authors' experience it is the main cause of error when using pipettes. The technique can be practised by drawing up volumes of distilled water and then allowing them to run into a beaker so that the contents can be weighed, thus checking for consistency and accuracy.

The other common piece of volumetric glassware is the burette. This, in general terms, is less accurate than a pipette and usually would not be used to dispense a specified volume repeatedly. Its major use is in volumetric analysis where some unknown quantity of liquid is required. The burette is graduated along its length in cm^3 so, by taking readings before the required volume is run out and after, the exact volume run out can be calculated.

Burettes need careful cleaning after use, and if necessary the tap should have a slight trace of vaseline wiped on it before it is put away. It is a common mistake to overgrease the tap and a useful sequence to follow is to put a little on, and then wipe it off again; this leaves a small trace on the tap which is quite sufficeint. To facilitate reading which graduation corresponds to the bottom of the meniscus, it is often convenient to slip a piece of card, one half of which is coloured, the other white, over the end of the burette as shown in figure 8.9. The graduations on many burettes are at 0.1 cm^3 intervals, and with practice it should be possible to say between which two divisions the bottom of the meniscus falls.

Assume that in a certain experiment the liquid level is some-

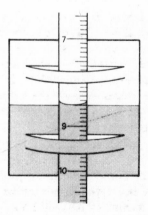

Figure 8.9

where between the 15.5 and 15.6 cm³ marks, a volume is run out, and the meniscus then is between the 45.5 and 45.6 cm³ marks. These two levels could be expressed as 15.55 ± 0.05 cm³ and 45.55 ± 0.05 cm³. The maximum volume that could have been delivered is $45.6 - 15.5$ which is 30.1 cm³ whereas the minimum volume could be $45.5 - 15.6$ which is 29.9 cm³. You can see that the volume delivered is somewhere in the range 29.9 to 30.1 cm³ which can be expressed as 30.0 ± 0.1 cm³. It is important to note that although the possible error in the reading of each level was only 0.05 cm³, the possible error in the delivered volume is twice that, that is, 0.1 cm³.

8.6 MULTIMETERS

A multimeter is a multi-purpose electrical measuring instrument. A typical example will have switches to convert it into a meter suitable for either alternating current (a.c.) or direct current (d.c.) and to convert it into an ammeter, a voltmeter or a meter to measure the resistance of a circuit or circuit element. In recent years the price of electronic instruments has fallen to the point where they are comparable in price to the traditional analogue instrument and they have the advantage of a simple-to-read digital output which, in most cases, places the decimal point in the right place so removing confusion in reading and recording the output. However, since analogue instruments are still widely used in industry and research it is appropriate to describe them in more detail.

A typical moving-coil or analogue meter is basically a microammeter; switches allow different resistors to be connected in series or in parallel with this meter which convert it into either a voltmeter or ammeter (figure 8.10). For resistance measurements, a small battery is connected in series with the meter and the two terminals of the instrument. If an external circuit is made, the lower the resistance between the terminals, the greater will be the current recorded, whereas the higher the resistance, the lower the current. Two points must be noted particularly in connection with the scales; the first is that while the current and voltage scales increase from left to right, that is, zero current is with the pointer on the left,

Figure 8.10

the resistance scale decreases from left to right, that is, zero resistance is with the pointer on the right. The second point is that the current and voltage scales are linear, which means that equal divisions along the scale correspond to equal amounts of current. The resistance scale is not linear, and considerable care must be taken when reading from it.

Measurements of current and potential difference are simply accomplished. First, the instrument must be set to either a.c. or d.c. as appropriate, and the correct range selected. If you do not know for certain which is the correct range, choose one that is too high, which makes the meter less sensitive; do not rely on a built-in cut-out or fuse to protect the meter. Then connect the meter in series for measuring current or in parallel for measuring potential difference (voltage). If there is only a small deflection of the pointer, then the range switch may be altered until a deflection between half and full scale is achieved. Then, for example, if the range switch is set to 12 V, a half-scale deflection would indicate 6 V.

Before making any resistance measurements, it is necessary first to check the zero of the instrument. The manufacturer will always supply instructions as to how this should be done on his particular instrument, but essentially the process will consist of connecting the two terminals directly together and adjusting a control knob until the pointer indicates zero resistance. The connection between the two terminals can now be removed and the instrument will give a direct reading of the resistance of any resistor connected between those two terminals, but remember the scale will not be linear, and care must be taken.

A simple use of a multimeter on its resistance range is to check for continuity, that is, to check that the circuit is complete. It may be necessary, for example, to check to see if a cartridge fuse has blown or not. If it is connected between the terminals of the meter and it is intact, there will be a circuit through it of very low resistance, and the meter should read zero. If the pointer stays at the other end of the scale, indicating a very high resistance, it is a good sign that the fuse has indeed blown.

8.7 MICROSCOPES

There are many types of microscope in use, including those which use X-rays, electron beams, ultraviolet radiation, etc., and some which use polarised light. We shall confine ourselves to those which use ordinary light, sometimes passed through the sample (transmitted light) (figure 8.11) and sometimes reflected from the top of the sample (reflected light) (figure 8.12) before it passes through the lens system.

Transmitted-light types are perhaps most commonly used in biology laboratories; they often have a mirror at the base which is used to reflect light from an external lamp up through the sample which is held, usually on a slide, on the microscope stage. The barrel of the microscope containing the lenses is free to move up and down relative to the stage and it is by this that focusing is achieved. When first using a microscope it is a good idea to rack the barrel down until the lower lens is just above the slide (a common fault of beginners is to force this lens down into the slide with consequent damage to both), then as you look through the microscope, the barrel can be raised slowly until the object can be seen clearly in focus. If the image is rather dark, then it is likely that

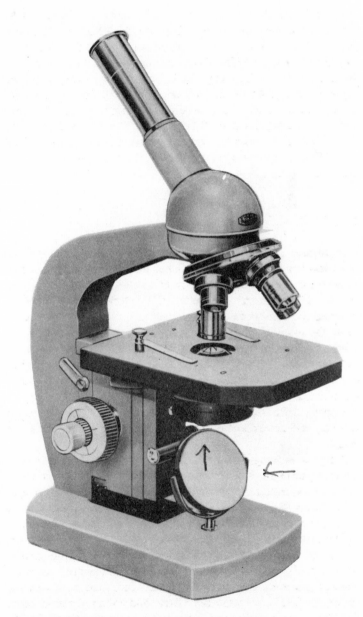

Figure 8.11 Microscope using transmitted light

the mirror at the base of the instrument is at the wrong angle, and this can be adjusted to give a clear bright image.

Reflected-light microscopes have a similar lens system to transmitted-light types, but do not have the base mirror. Instead a light bulb is included in the barrel together with a semi-silvered mirror so that light is reflected down the barrel on to the object and from the object back up the barrel, through the lens system to the eye.

Apart from a few routine operations, the servicing of a microscope should be left to an expert, because inexpert handling can leave complex lens systems in a worse state than they were found. The jobs that can easily be done include removing and cleaning the eyepiece with a soft fluffless cloth, gently cleaning the outside surface of the objective lens (the bottom one) with a soft cloth if it is very dirty or just dusting it off with a soft brush. The mirror can easily be cleaned with a proprietary window cleaner. The rack-and-pinion mechanism which moves the barrel up and down also frequently needs to be cleaned and given the thinnest smear of thin oil to remove any chance of a rough movement which makes accurate focusing almost impossible. It should not be allowed to work too loose or the barrel may be able to fall under its own weight with subsequent damage to the objective lens. A set of watchmaker's screwdrivers will make these jobs easier and minimise the risk of burring the heads of the screws.

8.8 SCIENTIFIC REPORTING

Whenever an experiment is performed, it is essential that a record is kept. There are several different purposes for which this record may be needed. First it serves as a reminder for the experimenter of what he did, which he may want to refer to some time later. (Records of coursework experiments at college should be useful for revision.) Secondly, it explains to others what was done and what results were obtained; public health analyses of suspected food or quality control checks on production are examples of this use. Thirdly, in research work, an accurate record is needed to ensure that the original work was done carefully and to explain to others what was done in order to avoid wasteful repetition.

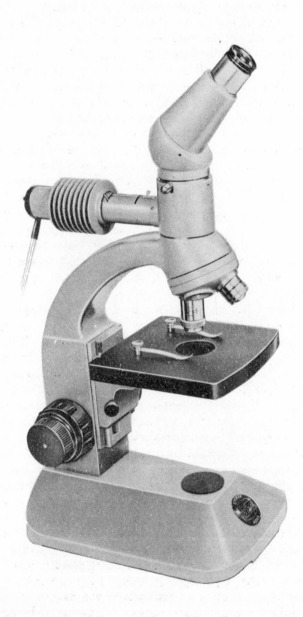

Figure 8.12 Microscope using reflected light

It often happens that the experimenter expects a particular result even before he starts the practical investigation. (This probably applies to many experiments performed at college.) It is tempting for a quality control technician to report that everything is satisfactory, rather than have to report faults which will disrupt production and may be the cause of losing bonuses. Even eminent research workers often perform experiments to 'verify' a theory which they or their colleagues have worked out. It is important always to be as clear, accurate and honest as possible in the reporting of experimental observations. The discovery of the inert gases, argon, neon, etc., in the atmosphere was delayed for several years because residual gas bubbles were dismissed as being due to imperfections in the removal of oxygen, nitrogen and carbon dioxide from dry air. There have also been examples of deliberate attempts to mislead, such as the Piltdown Man and some of the work on inherited intelligence. It is essential that evidence is not 'bent' in order to make the results fit some preconceived theory. Quite apart from the obvious consequences of being found out, 'rigging' may well obscure an important and interesting issue. Descriptions and measurements which are not the whole truth may well be a source of personal danger as well as financial loss.

Observations, particularly of numerical data, should always be recorded at the time that the observation is made. If you have to take a sample to an instrument to perform some measurement, take your notebook with you. Write down exactly what is being measured, the units in which it is being measured and an estimate of the accuracy of that reading. For example write 'the mass of sample $A = 21.76 \pm 0.01$ g', then label the object A; you will probably need to record the mass of many such objects before the day is finished. If you use a notebook for this, the right-hand pages may be used for such readings and the left-hand pages for sketches, diagrams, etc. You must also note any special precautions, events or unexpected difficulties. It is essential to use a notebook; separate pieces of paper are too easily lost or mixed up. If you are trying to develop a new procedure, it is obviously particularly important that you record each step as you take it. If you are successful, you may forget the exact sequence that led to that success.

Where there is a situation that requires a series of measurements or observations to be repeated many times on a large number of samples, it is usually worth while devising a standard form with spaces for each measurement and a final result or conclusion. In some laboratories where the bulk of the work is of a routine nature, such forms will probably be printed in bulk by the laboratory manager. In such cases, it is advisable that each form be given a sequential number and a carbon copy (or photocopy) kept on file by the worker who performed the tests.

In a research situation, you are likely to be asked to write a report on an experiment or investigation that you have performed, and this is certain to be asked of you in connection with some of the work that you do at college. It is often difficult to decide how much detail should be included in a report of this nature and how much may be left out. A useful guide is to find out whether a colleague with about the same technical knowledge as yourself could repeat your work with only your report to guide him. Would he avoid any pitfalls that caused you difficulty? Would he get results similar to yours? If not, you should suspect that your report is inadequate.

In addition to accuracy and honesty in compiling a report, you must be easily understood. It is important to use unambiguous language, and to give diagrams whenever these will be of help; the experimental results should also be presented in a way that makes them easily understood. Very often this will mean presenting your data in graphical or pictorial form. For example, you may be asked to investigate experimentally the way in which two variables depend on each other. A theoretical treatment may suggest that if property y doubles, then property x increases by a factor of four; if y triples, then x increases by a factor of nine. Expressing this mathematically, one would say that it is believed that x is proportional to y^2. The experiment would consist of setting the value of x and measuring the value of y; this would be repeated several times for different values of x.

You could present your results graphically by drawing a graph of x against y; if you do, however, it will be a curve (figure 8.13). If the theory was correct, then you would indeed expect a curve, but a special sort called a parabola. It is very nearly impossible to say whether any given curve is a parabola or not, so from the graph that you have drawn it would be almost impossible to say whether

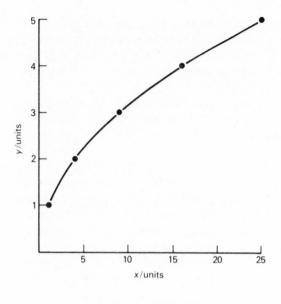

y/units	x/units
1	1
2	4
3	9
4	16
5	25

Figure 8.13

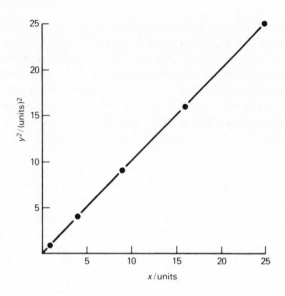

y/units	x/units	y^2/(units)^2
1	1	1
2	4	4
3	9	9
4	16	16
5	25	25

Figure 8.14

the theory was correct or not. An alternative graph is shown in figure 8.14; here the values of x are plotted against the squares of the values of y. If the theory is correct, and x is proportional to y^2, then you should get a straight line. It is easy to recognise whether a line is straight or curved, and, in this example, a straight line was obtained so giving experimental confirmation of the truth of this particular theory over the range of this experiment.

Although it is a good idea to rearrange things to give a straight line if possible, do not assume that all graphs you draw should be straight lines. Most graphs will, however, be smooth curves. If the results that you obtain do not give points lying on a smooth curve, then draw the best curve that you can, passing above some points and about equally below others. In general, draw smooth curves and not a series of short straight lines; graphs such as these are only used in special cases such as instrument calibration graphs mentioned elsewhere in this book.

Some data are more simply presented in pictorial form rather than graphically. For example, quantities observed in particular places or within stated times are usually best presented in the form of histograms (figure 8.15). Where you wish to show how a total collection was broken down into subunits, a pie chart may be the most useful form in which the information could be presented (figure 8.16).

In an environmental investigation, it is required to note how many motor vehicles pass a given point (in either direction) in each hour of the working day, the information being collected by two people standing at the road side. This data is best represented in the form of a histogram.

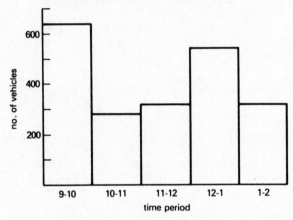

time interval	vehicles
9.00-10.00	640
10.00-11.00	280
11.00-12.00	320
12.00-1.00	540
1.00-2.00	320

Figure 8.15

It is required to show what happened to a selection of workers in a large laboratory complex with a high staff turnover rate. One hundred people were chosen who were employed in this laboratory, on 1 April 1977, and their occupations on 1 April 1978 discovered. The results of a survey such as this are most easily understood if they are presented in the form of a pie chart.

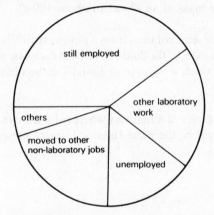

The angle at the centre of the circle due to each segment is proportional to the number of people in that category.

occupation in 1978	no. of people
still employed	40
other laboratory work	20
unemployed	15
moved to other jobs (non-laboratory work)	20
others	5

Figure 8.16

8.9 ASSIGNMENTS

8.1 Use (a) a pair of vernier callipers, (b) a ruler and (c) a micrometer screw gauge to measure a variety of distances, record your observations and the accuracy of these readings.

8.2 Of the various types of balance available to you at work, which is the most accurate? To what percentage accuracy could it determine the mass of an object of about 100 g?

8.3 Run out several volumes from a pipette into different beakers, determine the mass of the fluid delivered in each case and assess the accuracy to which a volume of liquid can be obtained using a pipette.

8.4 Give examples of situations where (a) a histogram and (b) a pie chart would be the most suitable form of presenting experimental results.

9 Laboratory Standards

9.1 MEASUREMENT

In our study of science, we try to make measurements and use these measurements to help us gain an understanding of the real world. In everyday life, we frequently have to make estimates of times, distances, etc., to decide, for example, whether we can catch up and board a moving bus, or whether the car that we are driving will (or will not) pass through the available gap. Usually these estimates are correct; on the other hand, if people are asked to state the height of someone they have witnessed, or state distances after a road accident, the values estimated are usually wildly inaccurate.

The ease of deceiving the human eye can be demonstrated by any of the well-known optical illusions. Consider figure 9.1. Either cut out the small arrow shown in figure 9.1b or draw a similar arrow on a small piece of paper. Place it on the line XY of figure 9.1a so that it looks something like figure 9.1c. Move the small

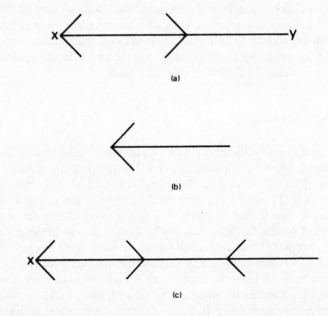

(a)

(b)

(c)

Figure 9.1

arrow backwards and forwards until the two parts of the centre line appear equal. Now measure these two distances with a ruler. Although they looked equal, it is quite usual for one to be about three-quarters of the size of the other. This shows how unreliable the human eye can be in estimating distances.

All processes of measurement, whether of length, time interval or pH of a solution, are basically processes of comparison. The thing being measured is compared with something whose size is already known. In simple terms, if the height of a man is twice the size of a piece of wood, and it is known that the piece of wood is one metre long, then the man is two metres tall; but how is the length of the piece of wood known?

The UK standard of length used to be the yard, which was defined as being the length between two marks on one particular bar of metal in London; similarly the metric standard of length, the metre, was fixed by a metal bar in Paris. In principle, anyone who wanted to measure a distance went with a piece of wood or metal to either London or Paris and cut it to the same length as the standard. The original metal bars were called 'primary standards', and the copies that were made by direct comparison were called 'secondary standards'. Manufacturers of rulers might have owned such a secondary standard and used it as the unit of length against which to check all the rulers that they made.

9.2 STANDARDS

The first major difficulty is that of accurately maintaining the standard. Most metals expand when their temperature is raised, and so the standard yard had to be kept at a constant temperature, nor was it allowed to sag between its supports, and so on. It is apparent that a bar of metal was not a good choice for the standard of length. What was really wanted was something, or some property, that remained constant over wide ranges of temperature, pressure and humidity and did not alter over a period of hundreds of years at least. Once a ruler manufacturer had obtained his secondary standard, he, as well, had to take steps to preserve the original size. If, for example, it was dropped, it could be distorted and cease to be a useful standard.

A second difficulty is one that is well known to technicians. Suppose you were asked to cut twenty pieces of wood to a specified size. You might have measured the first one, then used that to measure the second, the second to measure the third and so on. Any error you may be making would steadily become amplified, producing a relatively large error at the end. An identical process takes place in the 'standardisation' of the less expensive instruments. Very few manufacturers would keep a primary standard, and not many keep a secondary standard. Probably whatever they were using as a standard had at some stage been compared with a secondary standard, and they then compared their manufactured articles with this. This is obviously a repetition of the case of the pieces of wood.

It can be seen that any property that is used in laboratory standardisations must, first and foremost, be one that will remain unchanged over a long period of time. It must remain constant irrespective of the physical environment surrounding it, and it must be possible for a suitably equipped laboratory, anywhere in the world, to reproduce it exactly. Without this latter condition, it would be impossible for scientists and technologists to exchange experimental results and measurements easily and meaningfully and the import and export trading of all nations would be unduly complicated. It is bad enough having (at the time of writing) two different systems of units in the United Kingdom and the rest of Europe, but imagine the chaos if each country used the centimetre as a unit of measurement, but this was differently defined in each country. A one centimetre nut made in France would not fit a one centimetre bolt on a German car. When the United Kingdom goes over to the standardised metric system, it should be possible for a whole range of fittings to be common to all cars manufactured in Europe.

The SI system of units is an internationally agreed system which defines seven basic quantities. The standards that are used for these quantities all fulfil the conditions laid out above and these form the basis of modern scientific measurement.

Standards have been chosen for the following quantities

| metre | (m) | to measure length |
| kilogram | (kg) | to measure mass |

second	(s)	to measure time
kelvin	(K)	to measure temperature
ampere	(A)	to measure electric current
candela	(cd)	to measure luminous intensity
mole	(mol)	to measure the amount of a substance.

The experiments that are necessary to set up a primary standard of any of these quantities are costly and difficult and in most cases require highly specialised equipment. In most countries, governments have established laboratories whose main work is to provide these primary standards as and when they are required for the calibration of suitable secondaries. In the United Kingdom we have the National Physical Laboratory at Teddington, which is also responsible for research into maintaining and improving these standards and the methods of realising them.

Although these standards exist, it is generally beyond the scope of most small research or industrial laboratories to attempt to compare their instruments with primary standards that they could establish for themselves. They may only possess secondary, or even lower, standards.

Specialised laboratories do exist, on a commercial basis, to standardise instruments and they will normally make comparisons with secondary standards.

9.3 THE USE OF STANDARDS

In the day-to-day work of a laboratory, it is frequently necessary to establish standards. For example, a technician might need to know the strength of a particular sample of hydrochloric acid. There are several different ways in which this might be done. Indicator papers change colour according to acidity and, by comparing the colour of the paper with colours on a chart, the strength of the acid could be determined, although only approximately. Of course, the method would be of little use to a technician who is colour blind.

As an alternative, a pH meter could be used. In essence this is an electronic instrument in which a pointer moves across a scale when a probe is placed in the liquid under test. The meter reading is related to the pH of the sample and this in turn can be related to the strength or molarity of the acid. The problem is to know to what value of pH a particular scale reading corresponds. Although all such meters have a scale that is directly calibrated in pH values, electronic components change with age and temperature and just because the meter was correct when it left the factory certainly doesn't mean that it is still correct—it could be indicating a pH value of 4 whereas the true value is 5 or 6. BS 1647:1961 states, that the primary standard of pH is a one-twentieth molar solution of pure potassium hydrogen phthalate at 15 °C. This is defined as having a pH of exactly 4. In order to check the operation of the pH meter, it would be necessary to prepare accurately for yourself a standard solution made up to this specification.

The normal way of preparing such a solution would be to weigh out accurately the required quantity of the chemical and carefully dissolve it in a volumetric flask. The accuracy of the standard made in this way will depend on the accuracy not only of the balance but also the flask used and the care with which the operation is carried out. The accuracy of these pieces of equipment would have been checked by the manufacturer by comparing them against the standards that he was using. Manufacturers usually issue certificates specifying the accuracy of pieces of equipment that they have checked particularly carefully, and even more careful checking may be carried out, for a fee, by the National Physical Laboratory.

In this case it was necessary to create a standard with which to compare. This is very frequently the case in the chemical and biological sciences. Had it been necessary to determine the potential difference between two points in a circuit, a voltmeter could have been selected, either an electronic type or a moving-coil instrument. This meter could be connected across the two points in question and the scale reading noted. Again the same question must be asked: how can the user be confident that the meter is indicating the correct value? Someone might have altered the zero setting, or have been using the meter for some specialised application that involved internal modifications, or it might have been dropped at some time in its life and the manufacturer's calibration upset. The only way in which you can be certain is to check the voltmeter periodically against a suitable standard.

Weston standard cells are readily available and at a stated temperature these produce an accurately reproducible voltage. This voltage, written on the cell casing or printed on an accompanying certificate, is usually of the order of 1.018 V. There are certain limitations on the use of these cells, but in principle, one, then two, then three in series could be connected to the voltmeter and the scale reading compared with the actual applied voltage. If these values agreed within the accuracy to which the meter could be read, it would be reasonable to assume that it was accurate at points between the values that had been checked (In practice it is unwise to standardise a moving-coil voltmeter directly with Weston cells due to the risk of damaging them by drawing too large a current. The correct method involves the use of a potentiometer, and is described in physics textbooks.)

In this example, rather than having to make up a standard to check an instrument, a suitable standard is cheaply available and this may be used to check the calibration of the instrument.

We would not wish to give the impression that all laboratory instruments should be viewed with mistrust and suspicion. The great majority of manufacturers go to considerable lengths to produce equipment that is reliable and will give accurate repeatable readings over a period of many years. However, it must be borne in mind that accidents can occur which might upset a calibration. When an instrument has seen many years of service, its moving parts may be worn or, in the case of an electronic circuit, components will have aged and may no longer be up to their original design standards. Finally it must be remembered that few instruments are completely user-proof; continuous overloading of an instrument, for example, may well upset its original calibration.

Except for very specialised or sensitive work, it is quite unnecessary to check the calibration of an instrument every time it is used but, none the less, this is a task that should be carried out at regular intervals in much the same way that a car needs regular servicing. The frequency of such checks will depend on many factors including frequency of use and robustness of construction. A competent technician will devise a programme whereby all the instruments under his care will be regularly cleaned and checked in the ways outlined above.

It is not within the scope of this book to list in detail the methods that could be used to calibrate even a fraction of the devices to be found in a modern laboratory. Many of these are highly specialised and it will be necessary to follow carefully the instructions in the handbook supplied by the manufacturer. The principles whereby such checks can be made, however, are almost universally applicable. Obtain several standards of the quantity that you are trying to measure and apply these to the instrument in question. Then, either adjust the instrument so that its scale reading is equal to the value of the standard that has been applied, or, if this is not possible, note the indicated value and the true value and plot a calibration curve.

Example

Calibration of an Electronic Voltmeter

The standards to be used in this case are several Weston cells. Calibration certificates are provided with these cells stating that, as long as they are not connected to a resistance lower than one megohm and are in the temperature range 15 to 25 °C, they will provide the stated voltage. Because electronic voltmeters have a very high resistance, usually several million ohms, Weston cells may be connected to them; the precautions mentioned earlier for standardising moving-coil voltmeters are therefore less important here. Two calibration controls are provided on the voltmeter, both adjusted by set screws on the case. The first is a zero control, the second a gain control which controls the gain of the input amplifier; if this is increased the meter will read a higher value, and if it is decreased the meter will read a lower value, even though the same voltage is connected across the terminals in each case.

First, a short piece of wire can be connected across the voltmeter terminals. The applied voltage is now zero and the meter can be adjusted until it shows a zero reading. After the piece of wire is removed, one of the standard cells can be connected to the meter and the meter gain control adjusted so that the meter indicates exactly the voltage available from the standard cell. Now two cells can be connected in series to the terminals and the meter should now read the sum of the voltages available from the cells; if it does not, the meter is said to have a non-linear response and the actual value shown must be noted. This procedure can be repeated for combinations of as many cells as are available, which will check the calibration up to the maximum voltage obtainable, and the results may be represented graphically as shown in figure 9.2.

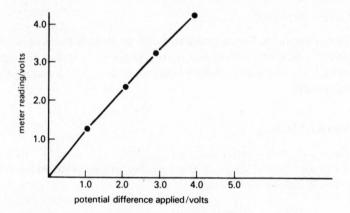

Figure 9.2   Meter calibration graph

In many cases this calibration curve is drawn as shown in figure 9.3, where the indicated value is drawn along the horizontal axis and the correction plotted vertically. If, for example, the meter indicates 4.0 V, the correction is + 0.1 V, so the true value is 4.1 V.

Note that this is one of the few occasions where it is quite correct to join the points on the graph with a series of straight lines. This is done because we have no way of knowing exactly how accurate the meter is between the points at which we have checked it.

The reason for the phrase 'non-linear response' is now obvious. When the meter is next used, the graph can be used to determine real values from indicated values read during the course of an experiment.

Note 1

Weston cells must not be connected to any meter other than a

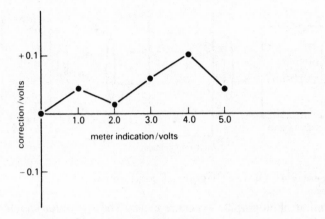

Figure 9.3   Meter calibration graph

high-impedance electronic voltmeter such as the types having a FET input stage.

Note 2

While the above is a useful illustration, any meter with such a dramatically non-linear response has a serious fault and should be placed on one side for repair with a note specifying why you have done this. With such a note, the service engineer need not waste his time repeating the checks that you have just made.

Frequently, commercially available meters that are calibrated in terms of the quantity to be measured will not be available, but provided that the property in question causes some sort of effect, then the sort of technique outlined above can be used to produce a specifically made 'once-off' calibrated instrument.

Example

It is required that the rate at which a chemical reaction proceeds be observed. It is observed that the reaction goes relatively slowly and as more and more product is made the colour of the solution gets deeper. A simple experiment could be set up such as that illustrated schematically in figure 9.4. The photoelectric cell is like the

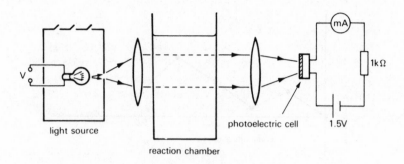

Figure 9.4 Outline diagram of a colorimeter

familiar photographic exposure meter. The apparatus should be shielded so that the only light falling on the cell comes from the source through the reaction chamber.

Solutions of the coloured products, of known concentrations, could be introduced in place of the reactants and the corresponding meter readings noted. Provided that the dimensions of the containers of both the reactants and the standards are kept the same, it will be possible to produce a calibration graph showing meter reading against the concentration of the coloured products

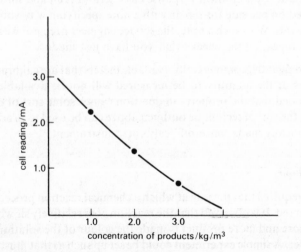

Figure 9.5 Colorimeter calibration curve

(figure 9.5). With the help of this graph, the course of the original reaction can easily be followed.

9.4 COMMONLY AVAILABLE STANDARDS

9.4.1 Physics Laboratories

Voltage Reference

This is usually a Weston standard cell, as already mentioned. It gives an accurate voltage but is susceptible to overload damage and must not be used to deliver a current in excess of a fraction of a microamp.

Standard Masses

These are immediately familiar as 'boxes of weights'. Every laboratory should keep one good-quality and accurate set for the periodic checking of balances.

Standard Lamps

These are white light sources (approximating to illuminant A — CIE publication No. 15, 1971) and are usually tungsten filament bulbs in a large glass gas-filled envelope. They have usually been calibrated against a secondary standard to give a known light output when there is a stated current flowing through the filament with a known voltage across its ends. They may be used for the calibration of photocells and in other light intensity measurements.

Standard Radioactive Sources

Any laboratory dealing with radioactivity will, at some stage, need to standardise equipment, not only in terms of source strength, but also the energy of the radiations that are being detected.

The Radiochemical Centre at Amersham will supply sources of a known activity at a stated time, or, for example, gamma sources that emit gamma radiation of specified energies.

9.4.2 Chemistry Laboratories

One of the chemist's most powerful tools is spectroscopy, where measurements are made of the wavelengths of the light emitted from or absorbed by different molecules or atoms. Some of the lines in the emission spectra of krypton, caesium and mercury, among others, have wavelengths which are very precisely known and these can be used to standardise instruments in terms of wavelength measurements.

Standard Solutions

Standard solutions can be made up by dissolving analytically pure material in a known volume of water (or other appropriate solvent). The requirements of these primary standards are that the materials should be easily obtained in a chemically pure state and that they should be stable in air at normal laboratory temperatures. It is also helpful if they are readily soluble. Such standard solutions are used in all aspects of quantitative analysis.

Conductivity Cell Calibration

Standard solutions are defined for use in calibrating conductivity cells. The conductivity of potassium chloride solutions of known concentration and at stated temperatures are precisely defined.

pH Measurement

BS 1647: 1961 lists solutions that may be used for the calibration of the glass electrodes used in pH measurement.

This is not intended to be an exhaustive list of all the standards that are available but merely a selection. Specialist books on physics, chemistry and biology usually refer to the relevant standards as they come to that topic of work, and technicians need only be familiar with those that are of immediate practical application to their work.

9.5 THE BRITISH STANDARDS INSTITUTION

The BSI was formed in 1901 and was then known as the British Engineering Standards Association. It publishes standards giving the sizes and names of units to be used in the United Kingdom and also standards to be followed in the manufacture of a great variety of articles. There are well over 3000 standards in current use and they are identified by the letters BS followed by a serial number and a publication date. For example

> BS 3763:1970 deals with the SI system of units; it was published in 1970
> BS 5378:1976 specifies safety colours for use in industry; it was published in 1976.

In the United States there is an equivalent organisation, called the American Standards Association. It was founded in 1917 and performs a very similar function to the BSI. The most common ASA standard to be used in the United Kingdom is probably the rating that they have developed for film speeds. For example, a popular colour slide film is rated at 25 ASA, and the faster, or more sensitive, colour print film at 80 ASA.

Full details of all the British Standards that are available are given in the *British Standards Year Book*, a copy of which is usually held by all public and private technical libraries.

9.6 ASSIGNMENTS

9.1 What standards are available in the laboratory in which you work?

9.2 Are these primary standards, or have they been calibrated against a primary?

9.3 Taking some of the more commonly used measuring systems, balances perhaps, suggest how frequently these should be checked against standards.

9.4 Devise a schedule for checking regularly the major items of measuring equipment with which you come into contact.

9.5 Find out the titles and reference numbers of the BS publications that most apply to your own area of work.

Appendix

Basic Units

Name	Symbol
metre	m
kilogram	kg
second	s
ampere	A
kelvin	K
candela	cd
mole	mol

Prefixes

Factor	Name	Symbol
10^{18}	exa	E
10^{15}	peta	P
10^{12}	tera	T
10^{9}	giga	G
10^{6}	mega	M
10^{3}	kilo	k
10^{-3}	milli	m
10^{-6}	micro	μ
10^{-9}	nano	n
10^{-12}	pico	p
10^{-15}	femto	f
10^{-18}	atto	a

GENERAL PHYSICAL CONSTANTS

Quantity	Usual Symbol	Magnitude and Units
Speed of light (vacuum)	c	2.998×10^8 m/s
Atmospheric pressure (standard)	p	1.013×10^5 Pa or N/m^2
Planck constant	h	6.625×10^{-34} J s
Mass of electron	m_e	9.109×10^{-31} kg
Mass of proton	m_p	1.673×10^{-27} kg
Mass of neutron	m_n	1.675×10^{-27} kg
Charge on electron	e	1.602×10^{-19} C
Avogadro constant	N_A	6.022×10^{23} mol^{-1}
Boltzmann constant	k	1.381×10^{-23} J/K

RESISTIVITIES OF SOME METALS

Name	Resistivity $(10^{-8}\ \Omega\ m)$
Aluminium	2.7
Chromium	13
Copper	1.72
Iron	10.5
Lead	21
Nickel	7.8
Tin	11.5
Brass	6
Constantan (Eureka)	45
Manganin	43

THE E.M.F.s OF SOME CELLS

Daniell cell	1.06 V
Leclanché cell (dry battery)	1.53 V
Weston standard cell	1.018 V
Lead acid accumulator cell	2 to 2.15 V
NiFe accumulator cell	1.2 to 1.3 V

PHYSICAL CONSTANTS OF SOME METALS

Name	Symbol	Relative Atomic Mass	Density (10^3 kg/m^3)	Linear Expansivity (10^{-5} K^{-1})	Specific Heat Capacity (10^2 J/kg K)	Melting Point $(^{\circ}\text{C})$	Young's Modulus (10^{10} Pa)
Aluminium	Al	27	2.7	2.4	9.0	660	7.0
Chromium	Cr	52	7.1	0.85	4.5	1900	2.5
Copper	Cu	63.5	8.9	1.6	3.9	1080	12
Iron	Fe	56	7.9	1.2	4.5	1540	22
Lead	Pb	207	11.3	2.9	1.3	330	1.5
Nickel	Ni	59	8.9	1.3	4.5	1460	20
Tin	Sn	119	7.3	2.7	2.3	230	5.5
Steel	0.1% C; 12% Cr		7.8	1.0	4.6	1510	20
Brass	60% Cu; 40% Zn		8.5	2.1	3.8	900	10

THERMOELECTRIC POTENTIALS

If a junction with platinum is held at $100\,^{\circ}$C, and the other junction held at $0\,^{\circ}$C, then the following potentials measured in millivolts will be obtained.

Constantan	− 3.3
Nickel	− 1.5
Manganin	0.65
Copper	0.75
Iron	1.88

The e.m.f. developed by a copper–constantan thermocouple with one junction at $100\,^{\circ}$C and the other at $0\,^{\circ}$C can be calculated by looking at the materials above; the e.m.f. will be the difference between the quoted values for the two materials, that is, $0.75 - (-3.3)$ mV $= 4.05$ mV. Similarly, the e.m.f. developed by a manganin–copper thermocouple will be $0.75 - 0.65$ mV $= 0.1$ mV.

STANDARD WIRE GAUGES

SWG	Diameter (mm)
10	3.25
11	2.95
12	2.64
13	2.34
14	2.03
15	1.83
16	1.63
17	1.42
18	1.22
19	1.02
20	0.914
21	0.813
22	0.711
23	0.601
24	0.559
25	0.508
26	0.457
27	0.417
28	0.376
29	0.345
30	0.315

SCREW THREADS

Standard BA (British Association) threads are perhaps the most commonly used in science laboratories despite the reommendation in 1965 that industry adopt ISO metric threads; although BA threads will therefore eventually become obsolete, some details are given here.

BA No.	Diameter of Bolt (mm)
0	6.0
1	5.3
2	4.7
3	4.1
4	3.6
5	3.2
6	2.8
7	2.5
8	2.2
9	1.9
10	1.7

Details of other threads that may be encountered are published by the British Standards Institution

ISO metric threads, BS 3643:Part 1:1963
Whitworth form threads, BS 84:1956
Unified screw threads, BS 1580:1962 and 1965
further details on BA threads are in BS 93:1951

ANTIDOTES

Cyanide Poisoning

If cyanides have been swallowed, the antidote consists of two solutions, 'A' and 'B'. 50 cm³ of solution A is placed in a small polythene bottle, and labelled CYANIDE ANTIDOTE A; 50 cm³ of solution B is placed in a similar bottle and labelled CYANIDE ANTIDOTE B. Both bottles should bear the instruction 'Mix the whole contents of bottles A and B and swallow the mixture'.

Solution A is 158 g of ferrous sulphate and 3 g of citric acid crystal BP made up in 1 l of cold water. This solution deteriorates fairly rapidly and must be replaced regularly. Solution B is 60 g of anhydrous sodium carbonate made up in 1 l of water.

If hydrogen cyanide has been inhaled, the vapour of amyl nitrate (available in capsule form) should be inhaled.

In cases of cyanide poisoning, always summon a doctor as quickly as possible; he will be able to give further treatment by injection.

Phosphorus Burns

Wash the affected area with a 3 per cent solution of copper sulphate in water.

Bromine Burns

Wash copiously with water to remove all traces of bromine. Dilute ammonia or a 6 per cent sodium thiosulphate solution have been recommended as treatment. Obtain medical assistance as soon as possible.

Acid or Alkali Burns

Wash copiously with water; attempts to neutralise acids with dilute alkalis or vice versa are probably not worth while. Obtain medical assistance.

Phenol Burns

Wash copiously with water and then swab with glycerol. Obtain medical assistance as soon as possible.

SOME CARCINOGENIC SUBSTANCES

Alpha or beta-naphthylamine
Diphenyl substituted by at least one nitro or primary amine group; in addition, further substitution by a halogen, methyl or methoxy group, for example, o-toluidene

Nitrosamines
Nitrosophenols—except p-nitrosophenol
Nitronaphthalenes
Vinyl chloride monomer

Add to the list above those other known carcinogens that you may encounter at work, and those substances whose carcinogenic properties become known.

SAFETY COLOURS AND SAFETY SIGNS

Over the years, several different systems of signs have arisen. With increasing international trade and travel, standardisation is obviously desirable. BS 5378:1976 Safety Colours and Safety Signs is in agreement with EEC requirements and with most ISO (international) requirements; BS 5378 does not cover signs for transport of dangerous goods, traffic signs, fire safety signs or such signs as are used in cinemas, etc.

Colours

Red   used for Stop, Prohibition signs
       also for Fire fighting equipment and its location
Yellow used for Caution, Hazard warning signs
Green used for Safe conditions, Emergency exits and showers

Blue   First-aid and rescue stations
        used for Mandatory (obligation) signs, General
        information signs

Signs

Caution, toxic hazard
(black on yellow, black edge)

Figure A.3

Warning
Yellow background with black border. Safety symbol or wording
    in black.

Fire, open light and smoking prohibited
(black on white, red circle and bar)

Figure A.1

Prohibition
Red band and crossbar with white background. Safety symbol in
black.

First aid
(white on green)

Figure A.4

Information
Safety equipment: green background with white symbol or
    writing.
Fire fighting equipment: red background with white symbol or
    writing.
General information: blue background with white symbol or
    writing.

Respiratory protection must be worn
(white on blue)

Figure A.2

Obligation
Blue circle. Safety symbol in white.

Further Reading

First Aid

—*Digest of First Aid* (St John Ambulance Association and Brigade, London, 1970).

A. Ward Gardner and Peter J. Roylance, *New Essential First Aid* (Pan, London, 1972).

—*First Aid Manual* (St John Ambulance Brigade and British Red Cross Society, London, 1972).

Technicians and the Law

A. J. D. Cooke, *A Guide to Laboratory Law* (Butterworth, Borough Green, 1976).

Chemistry Laboratories

G. D. Muir, *Hazards in the Chemical Laboratory* (Chemical Society, London, 1977).

L. Bretherick, *Handbook of Reactive Chemical Hazards* (Butterworth, Borough Green, 1975).

Biology Laboratories

D. J. Short and D. P. Woodnott, *The I.A.T. Manual of Laboratory Animal Practice and Techniques* (Crosby Lockwood, London, 1969).

J. H. Seamer, *Safety in the Animal House* (Laboratory Animals, Huntingdon, 1972).

J. R. Norris and D. W. Ribbons (eds) *Methods in Microbiology*, 8 vols (Academic Press, London, 1969–73).

General

—*Safety in Science Laboratories*, D.E.S. Safety Series No. 2 (H.M.S.O., 1976).

Safety in Further Education, D.E.S. Safety Series No. 5 (H.M.S.O., 1976).

K. Everett and E. W. Jenkins, *A Safety Handbook for Science Teachers* (John Murray, London, 1977).

Various publications of
    Imperial College of Science and Technology
    Association for Science Education
    Department of Education and Science
    Health and Safety at Work Executive
    British Standards Institution
    Royal Society for the Prevention of Accidents